Computing Power Drives the Future

Computing Power Drives the Future

William Meisel

CONTENTS

Introduction

Computers have long done some things better than humans. They can do arithmetic faster and remember more information than we can.

We use digital technology constantly. We surf the Web. We store digital photos. We listen to streamed music and movies. We can't do without our smartphones.

Companies run their operations on specialized software. You trust your bank to keep track of your money.

Advances in technology have long delivered more computing power each year. A typical estimate is that more complex chips have doubled computing power every two years. Another way to look at that trend is that the cost of computing power has dropped by half every two years. That rapid advance has allowed digital systems to do more each year. Your smartphone is a small computer and does many of the things that required a personal computer years ago.

That computer power crossed a threshold when it made artificial intelligence feasible. The technology driving today's AI – deep neural networks – isn't new. The core methodology was known as far back as the 1980s, but computers were too slow to do the computation required. At some point, the cost dropped enough to enable the first successful applications. Today, computing power has reached a point where the large neural networks of generative AI impress us with their ability to write articles, create images, or write computer code at our command.

Computer power is increasing at an accelerating rate, driven by supercomputer centers using specialized chips such as Nvidia's graphics processing units. That raises the question of what comes next for AI and what other innovations this rapid growth in computer power will bring. It is likely to bring both benefits and challenges that society must deal with. Some experts even view AI as a threat to the human race.

When does computer power go beyond a tool to a threat? Today, most of us talk to our smartphones – speech recognition has long gone beyond the tipping point of usability. That was not an easy task. My previous book, *The Lost History of "Talking to Computers"*, documents the decades of struggle by hundreds of companies to jump that hurdle. Most people would consider speech recognition a part of AI, but I am not aware of it being attacked as an existential threat.

The rapid growth in computer power has a major economic impact. Companies building and maintaining AI software will have to continue making major R&D investments to take advantage of what that increased power allows. Companies not as directly impacted will see improvements in what their software can do, whether it is internally developed or acquired. The continuing improvements in cloud computing services such as Amazon's AWS means a company of any size can rent the most powerful computing power available without a capital expenditure.

Another major impact of increasing computer power is the need for increasing electrical power. At a time when climate change is a major challenge, growth of computer centers will at least temporarily increase carbon emissions.

This introduction is a quick summary of the important role computer power will play in our future with its unique ability to move from an idea to a solution much faster than previous innovations. The following chapters delve into the implications of this trend in depth.

2

Computer Power

The core trend this book addresses is the role of computers in economics and society. Many other innovations have had deep impacts on society. For example, the automobile changed the meaning of distance and created the oil industry. But most innovations have evolved slowly. Automobiles required roads. Electric vehicles required charging stations. Slow evolution allows timely addressing of problems created by an innovation, such as building pipelines to get oil from wells to refineries and controlling intersections with traffic lights. Even the Internet grew slowly; people needed connections. Computer power does require infrastructure—computer centers and electric power—but it is positioned to expand much faster than past breakthroughs, partly because of its exponential growth, but also because of the availability of capital from investors and large technology companies to fuel that growth.

Today, computer power has passed a 'tipping point' of capability and penetration of our everyday lives. The core digital technology had an early evolution from vacuum tubes to transistors to integrated circuits. Once the core computation was done by integrated circuits, Moore's Law took over. Gordon Moore, co-founder of Intel, predicted the number of transistors that would fit on a chip would double every two years, essentially saying that computers could compute twice as fast at a comparable cost every two years. That's exponential growth and, since the early 1970s, the number of transistors that fit on a chip has increased by a factor of *100 million*. We directly experience that growth when we upgrade our PCs or smartphones or visit a website. Businesses experience it every time they install a new server in their computer center.

And computer power growth does not benefit solely artificial intelligence, although the press recently seems to focus on that topic. For example, Elon Musk commented in a July 2025 webinar about AI that his company SpaceX uses by simulation of the *direct physics* of his rockets ("computational element analysis and fluid dynamics," both taking substantial computer power) to predict what will work before it is built. He said, "The test simulations are so accurate that if the test doesn't match, you assume the simulations were correct." He argued that building this kind of logic into AI would create future breakthroughs.

What is summarized here as growth in 'computer power' has multiple components:

- Moore's Law and advances in chip structure and size allow growth in the amount of computation possible *on a single chip*.
- There are *more chips* as companies add servers to their computing centers and consumers buy PCs and smartphones that do computation at the 'edge' of the network.
- Fairly recent innovations allow *parallel processing*, doing more than one instruction at a time on a chip, e.g. Nvidia's graphics processing units. They particularly increase the computing speed for certain types of processing (in particular, the neural networks of artificial intelligence).
- *Cloud computing services* such as Amazon Web Services and Microsoft Azure create huge computing centers that can take advantage of scale, allowing companies to use the most advanced technology over the Internet, charged on a usage basis.
- Demands for computing power to *develop ever-larger AI models* have increased the rate of investment in computer center expansion. Research Gate estimated that, after doubling every two years through 2012, computer power has been doubling *every 3.4 months*. In its July 25, 2005, issue, *The Economist* predicted, "By

2027 it should be possible to train a model using 1,000 times the computing resources that built GPT-4, which lies behind today's most popular chatbot." (This book uses the more conservative twice every two years in estimating computer power in a given year compared to 2025.)

- *Quantum computers* are at an early stage but promise a further major jump in computing power. In June 2025, IBM announced detailed plans to build an error-corrected quantum computer called Starling with significantly more computational capability than existing machines by 2028. The company hopes to make the computer available to users via the cloud by 2029. IBM claims Starling will be a leap forward in quantum computing, aiming for it to be the first large-scale machine to implement error correction. (Quantum computers are probabilistic, and can make errors.)

These factors are accelerating the current growth in computer power well beyond that predicted by Moore's Law. Some experts argue that we are approaching the limit of how small transistors on an integrated circuit can be, although *EE Times* noted in July 2025 that heat dissipation and interconnect delay, not transistor switching speed, have emerged as the primary bottlenecks. However, the other factors will sustain continued growth, particularly with some organizations planning huge computer centers. Some are so large they will be supported by their own nuclear power plant. For example, in June 2025, Meta signed a 20-year agreement to buy nuclear power from Constellation Energy, joining a growing list of tech giants turning to nuclear energy to meet the demands of artificial intelligence.

In January 2025, OpenAI announced Stargate, a new company formed to invest *$500 billion* over four years building new AI infrastructure for OpenAI in the United States. In July 2025, OpenAI signed a deal worth $30 billion a year to rent around 4.5 gigawatts of computing power from Oracle, a business-software giant, for Stargate. Oracle plans to build several data centers across America to meet OpenAI's demand for power.

In mid-2025, France's Mistral partnered with Nvidia on a data center powered by 18,000 cutting-edge chips. The EU announced plans for four 'AI gigafactories' with a $20 billion price tag. In June 2025, Meta unveiled a $14.3 billion investment and partnership that will be the core of a new artificial intelligence research lab dedicated to the pursuit of 'superintelligence,' essentially the goal of a computer system being smarter than humans. CEO Mark Zuckerberg reportedly said in an internal memo, "This will be the beginning of a new era for humanity."

Amazon is building one of the largest computer complexes for work with Anthropic, an AI startup, on 1,200 acres in Indiana. It is composed of seven data centers, each larger than a football stadium. The facility will consume 2.2 gigawatts of electricity—enough to power a million homes. It will use millions of gallons of water each year to keep the chips from overheating. Meta, which owns Facebook, is building a 2-gigawatt data center in Louisiana. OpenAI is erecting a 1.2-gigawatt facility in Texas and another in the United Arab Emirates. Despite these large computing centers being built to allow the optimization of very large neural networks, they are general-purpose computers that can be used for other purposes.

The 2020 book *How Innovation Works: And Why It Flourishes in Freedom* by Matt Ridley emphasizes that, historically, major innovations achieved wide use over a long period of incremental improvement. This has been the case for simple things like light bulbs to major innovations such as electricity. This slow movement from innovation to wide adoption is not the case with innovations driven by computer power.

For example, OpenAI launched generative AI pioneer ChatGPT, with little fanfare, in late November 2022. *MIT Technology Review* in 2023 quoted Sandhini Agarwal, who works on policy at OpenAI, as saying that in-house it was regarded as a 'research preview.' The publication quoted Liam Fedus, a scientist at OpenAI, as saying, "We didn't want to oversell it as a big fundamental advance." Yet, it very quickly became one of the most used applications on the Internet, reportedly reaching 100 million users in January 2023, just two months after launch. And in 2025, generative AI technology was being used by many companies, including by Google for summarizing the implications of many websites in a web search.

Most such services are continually being updated. Even more surprising, many generative AI services are free to consumers despite the significant expense of supporting them. With issues such as schools worrying about AI doing students' homework and a US government department publishing reports generated by AI (with errors generated by AI), the impact on society is already substantial.

The summaries provided by generative AI in response to requests have been challenged by providers of the core data used to build the models. In December 2023, the *New York Times* sued Microsoft and OpenAI, alleging that the tech companies owed billions of dollars for using copyrighted work to train ChatGPT. The newspaper's lawyers showed multiple examples of ChatGPT producing *New York Times* journalism word for word. They claimed this showed that AI tools do not substantially transform the material they're trained on, and therefore are not protected by the fair-use doctrine.

These claims had results. For example, in 2024, OpenAI signed a content agreement with News Corp, giving access to current and archived content from the *Wall Street Journal*, *Market Watch*, the *New York Times* and other publications. News Corp's five-year deal with OpenAI was reportedly worth over $250 million. The copyright issue is ongoing; in June 2025, Reddit, an online discussion forum, sued AI startup Anthropic for copyright infringement after reaching agreements for the use of their data with OpenAI and Google.

In June 2025, US District Judge William Alsup ruled that Anthropic's use of copyrighted books to train its AI model, Claude, qualified as 'fair use,' not a copyright violation. This decision addressed a lawsuit filed by three book authors who alleged that Anthropic infringed their copyrights by using pirated versions of their books for AI training.

Judge Alsup concluded that the AI training process was "exceedingly transformative," comparing it to a human writer learning from existing works to create new content. He said that the AI's purpose was not to replicate or replace the original works, but to generate distinct, innovative outputs. However, the court found that Anthropic infringed on copyrights by storing more than seven million pirated books in a centralized repository. This action was deemed not to fall under fair use, and a trial was scheduled for December 2025 to determine potential damages. Appeals are expected.

This brief history suggests how the rapid growth of computer power can drive innovations into broad use much faster than the historical norm. Such rapid impact on the economy and society represents challenges in controlling both positive and negative effects. I will address these challenges in later chapters.

Chips

The core of computer power is the chips (integrated circuits, semiconductors) that are its physical form. Chris Miller in the 2022 book *Chip War* said, "The idea that the semiconductor industry would eventually produce more transistors each day than there are cells in the human body was something the founders of Silicon Valley would have found inconceivable." As the complexity of these chips grows, their availability becomes a strategic issue for countries.

The production of chips has in part become a concern of governments because of an international supply chain. Two companies outside of the US and China are critical to making the world's most complex chips: Taiwan Semiconductor Manufacturing Company (TSMC) and the Dutch company Advanced Semiconductor Materials Lithography (ASML). ASML provides equipment that can take a chip design and project it on a chip with the narrowest of lines, and TSMC uses that equipment and more to produce the final chip. TSMC makes about 65% of the world's chips and almost all of the most complex. Miller noted the connection between the companies: "Both ASML and TSMC started as small firms on the periphery of the chip industry, but they grew together, forming a partnership without which advances in computing today would have ground to a halt."

Part of the motivation for the US to make more of the complex chips internally is the danger of China invading or blockading Taiwan. TSMC is investing heavily in Arizona, with plans for multiple fabrication facilities, including one that began mass production in late 2024. This investment, initially a $12 billion project in 2020, has grown to $65 billion, according to the National Association of Manufacturers.

China is behind in the chip war. In 2024, it spent a significant amount on importing chips: $385 billion, a 10.4% increase from the previous year. This is more than it spent on importing oil. The US has restricted sales of the most complex chips, creating problems for China. All advanced tech – from AI to missile systems, from automated vehicles to armed drones – requires cutting-edge chips.

3

Software and Algorithms

Computer power determines the speed at which software code can run. It can determine, for example, if, when you press a key to cause an action, such as turning on your PC, that action happens quickly or slowly. The speed of action depends in part on the hardware, the chips in your PC and their connections. But it also depends on the software, since the number of steps the computer must take before you can use your PC also controls how fast it starts up.

The word 'software' probably brings up a mental image of lines of software code. That is indeed an embodiment of software, the instructions that drive a digital hardware system. In addition, software typically uses data and creates data which one might think of as a table of numbers or text stored in some form of digital memory system. Today, almost all systems store data and the program in the same memory system—the 'von Neumann' architecture, first described in a 1945 report by John von Neumann, a multi-talented Hungarian and American mathematician. Computer systems have a central processing unit (CPU) that is the default hardware for the software.

All computer systems have a controlling layer: the operating system (OS). Operating systems such as Microsoft Windows and Apple macOS for personal computers, and Google's Android and Apple's iOS on smartphones do the heavy lifting in managing data in memory or hard drives, managing input (e.g. a keyboard and mouse) and output (e.g. putting information on a screen or playing an audio file), and many other functions. Apple lists six software platforms: iOS, iPadOS, macOS, watchOS, visionOS and tvOS.

Programs use the OS to run more specific applications such as Microsoft Word or a web browser without coding the basic functions the OS controls; software must be 'compatible' with the OS on the device. For example, Microsoft Word software differs for Windows and macOS.

A software program is a series of instructions directed at the processor that tells the processor how to deal with the data in memory. The core language of a computer is 'machine code.' This is the fundamental language that the computer's CPU can directly understand and execute. Machine code is essentially binary code, consisting of sequences of 0s and 1s. These binary digits represent instructions for the CPU, such as loading data into memory, performing calculations, or directing program flow.

Programming languages avoid the need to program in machine code by creating an easier way to define a procedure in a more intuitive form than 0s and 1s. Today, software is generally written ('coded') today in a 'high-level' language, in which commands are written in text, with names that suggest their function. A programmer can then review the flow of the program with an easier understanding of what it is doing. The high-level language is 'compiled' by software into machine code before being 'executed' by the CPU. Here is a basic example of a program that calculates and prints the area of a circle in the computer language Python, including descriptive comments by the programmer:

```
import math
```

```
# Define the radius of the circle
radius = 5
```

```
# Calculate the area using the formula: area = pi * r^2
area = math.pi * (radius ** 2)
```

 # Print the result

print(f"The area of a circle with radius {radius} is: {area}")

Programming languages have a very precise syntax. If a software program doesn't obey the grammar of the language, programmers will get an error message when they try to compile the program.

As the program is executed by the processor, it may change some data in memory that represents intermediate results. The software may be interactive with a user, responding to user input from a keyboard, a mouse, touch, speech, or other user interface, and delivering results through output devices such as a screen, speaker, or printer.

As software has become more complex, two factors have helped keep it from becoming completely indecipherable and unreliable. One is the use of modularity (e.g. 'subroutines' that perform a specific function). By breaking the software into manageable pieces, the programmer can more easily see what the program is doing, and can work on each module separately rather than one large program. When a module is useful in many programs, the programmer gets the benefit of reliable pre-existing modules without having to write them.

A second factor in the reliability of software is feedback from users. It's essentially impossible to test a program of any complexity for all possible conditions, so software almost universally has bugs in early versions. 'Alpha' and 'beta' versions of programs are usually issued in limited trials that uncover the most critical bugs. The final version of a software package will be tested through use by many more people in many more software and hardware environments, with resulting bug reports. Fortunately, bugs discovered by one user can be fixed by updates before all users encounter them. This feedback process is critical to the quality of today's software.

In an interesting observation that could broaden the definition of 'programming,' Brian Arthur in the 2009 book *The Nature of Technology* said, "Technology is a programming of phenomena to our purposes...The programming may not be obvious. And it need not be visible either, if we look at the technology from the outside." He is not talking of software code as the 'programming,' but he is looking at technology as an assemblage of components that is assembled according to some description in a 'grammar' (analogous to a software language) suited to that technology. For example, a circuit diagram describes how to assemble components on a circuit board to achieve a certain goal. A blueprint may describe how to assemble an aircraft. A programming language has a 'grammar' of allowable ways to write commands. Similarly, Arthur speaks of a technology domain: "A domain's grammar determines how its elements fit together and the conditions under which they fit together. It determines what 'works.' In this sense there are grammars of electronics, of hydraulics, and of genetic engineering."

Software code describes a series of steps that accomplish a goal. The goal may be relatively simple, such as computing an account balance in banking software. But, in many cases, the goal is much more complex. Before one writes software code, one must have methods of achieving the goal of the software, often characterized as an 'algorithm' when the method is sufficiently complex or mathematical. David Berlinski in 2001's *The Advent of the Algorithm: The Idea That Rules the World* defined an algorithm simply as "an effective procedure, a way of getting something done in a finite number of discrete steps." If an algorithm is defined as a 'procedure,' a computer isn't even needed; a written recipe is a form of algorithm with steps that a cook executes. But the simplicity of Berlinski's informal definition hides the importance of the concept now that computers can execute a huge number of discrete steps in a very small amount of time. In fact, he states: "The algorithm has come to occupy a central place in our imagination. It is the second great scientific idea of the West. There is no third." (The first great idea he refers to is calculus, the math behind much modern engineering and physics.) In practice, the term 'algorithm' usually implies the use of mathematical techniques for precisely and compactly stated goals. It is usually applied when the procedure is a particularly efficient way of accomplishing a task, perhaps motivating the use of the adjective 'effective' in Berlinski's short definition of an algorithm.

Often, an effective algorithm makes a task feasible that would otherwise be infeasible. To get a feel for the subjective term 'effective,' let's look at a simple example. If one has two numbers, a and b, and wants to compute aa + 2ab + bb, one could do so by the following steps: multiply a by a, a by b by 2, and b by b, and then add the results of the three multiplications. This requires four multiplications and two additions. However, if one realized that aa + 2ab + bb = (a+b)(a+b) by simple algebra, one could use the following algorithm: add a to b and multiply the result by itself, only one addition and one multiplication, an effective algorithm for computing aa + 2ab + bb.

This simple example is typical in that the usual objective of algorithms is finding ways to do certain procedures efficiently, and it suggests accurately that an efficient algorithm requires insight and invention. If a computation such as our simple example had to be done billions of times, efficiency becomes important to obtaining a quick result.

More complex algorithms can do things like finding the shortest-distance route between two locations on a map, even making it the shortest-time route if current traffic information is available. Richard Bellman's *Dynamic Programming* is a class of algorithms that can be applied to the shortest-route problem. They depend on what Bellman called the 'principle of optimality,' which asserts that the optimal solution to the overall problem is composed of optimal solutions to its subproblems. This states you can't improve the overall solution by making a suboptimal choice in any of the subproblems. If so, performance of the algorithm would suffer. This allows finding an optimal solution by breaking up the problem into a series of subproblems. In the shortest-route problem, it could be stated as "if you are anywhere on the shortest route, the rest of the route is the shortest route from where you are." This simple idea translates into a class of very efficient algorithms. People such as Bellman, who wrote 35 technical books (and inspired me as a teacher in graduate school and was a member of my PhD dissertation committee), are perhaps the equivalent in computer science to the pioneers in physics, but they certainly don't get the same recognition.

Algorithms can be implemented in software but usually start out as a description in text and equations. Books on standard computer algorithms may include chapters on esoteric subjects such as probabilistic analysis, sorting, data structures, graph algorithms, matrix methods, linear programming, dynamic programming and neural networks.

Another broad category of algorithm is of particular importance to the theme of computers connecting with people. These are empirical ('statistical') methods, where data is analyzed by software to extract patterns and produce a model predicting a likely outcome for given data describing a particular case. Those models are used to perform a task such as speech recognition.

Algorithms can recognize our speech and understand it well enough to do typical things we do with mobile phones, e.g. respond to "Give me directions to 123 Last Street," or "Text to Mary, 'I'm almost there.'" It can speak back to us in a synthetic voice generated by text rather than a recording, e.g. "Turn right at Green Street," using an algorithm that allows 'speech synthesis' from text.

Another example of an empirical method is optical character recognition (OCR), taking a 'picture' of text (e.g. from a page scanned into the computer) and turning it into searchable and editable text. The software for doing so is created by processing many examples of text to extract the characteristics that allow distinguishing letters. This type of process is called 'pattern recognition.' The methods used vary depending on the type of patterns being distinguished.

An algorithm that has shown the power of effective algorithms is machine learning using deep neural networks. It is driving today's artificial intelligence, as discussed in the following section. It is important to note that discovering 'deep learning' didn't result in its use; the methodology has been known for decades. Computer power had to reach the point where it was economically feasible. Thus, the practicality of an algorithm can depend on the state of computer power evolution.

4

Artificial Intelligence

John McCarthy, a Dartmouth professor in the 1950s, initially used the term 'automata studies' to describe the pursuit of machines capable of automatic behavior. When research on the subject didn't attract much attention, he used the term 'artificial intelligence' to describe it. It proved an effective marketing label.

But the term itself drove research toward trying to do with computers something that was previously limited to human intelligence. Early research focused on 'expert systems.' Expert-system software attempted to mimic human reasoning using a computer driven by a series of 'logic' instructions. The instructions were an attempt to react 'intelligently' to anything the user typed.

In 1958, MIT professor Joseph Weizenbaum invented an AI system called ELIZA that drew headlines. ELIZA was a chatbot that you could converse with through an electric typewriter hooked up to a large mainframe computer. ELIZA used simple rules to seem attentive and responsive to a user. One of the most popular versions of ELIZA was designed after a method of talk therapy that probed a patient by reflecting back what they said and gently asking for more information. In a paper Weizenbaum later published on the inner workings of ELIZA, he provided the following dialogue as an example:

User: Men are all alike.

ELIZA: IN WHAT WAY

User: They're always bugging us about something or other.

ELIZA: CAN YOU THINK OF A SPECIFIC EXAMPLE

User: Well, my boyfriend made me come here.

ELIZA: YOUR BOYFRIEND MADE YOU COME HERE

User: He says I'm depressed much of the time.

ELIZA: I AM SORRY TO HEAR YOU ARE DEPRESSED

User: It's true. I am unhappy.

ELIZA: DO YOU THINK COMING HERE WILL HELP YOU NOT TO BE UNHAPPY

Weizenbaum designed the system as an experiment. He wanted to see how easily humans might fall for an illusion of intelligence.

Expert-system technology never progressed to something truly comparable to human intelligence. The research basically collapsed into developing computer languages for describing the rules, but humans had to develop the rules. This was essentially writing a computer program in a specialized language, but it provided no guidance on what rules would be effective. There were some researchers trying to take a 'connectionist' approach, something that would lead to neural nets in the long run. But the limits of computer power blocked any real progress. The struggles led to failures which in turn led to 'AI winters,' periods when it was almost impossible to get funding for AI development and publish papers on AI.

The continuing increase in computer power has always driven change. When meaningful computer power could be packaged in a smaller box, computers became 'personal computers' (PCs). When it allowed a graphical user interface to operate quickly enough to create a point-and-click paradigm, computer use became more intuitive. Computer power allowed the World Wide Web to grow from a cute trick to what it is today. Smartphones became possible when enough computer power could be packaged in a small package. Computer games became increasingly realistic and addictive. Microprocessors make everything from televisions to microwave ovens easier to use.

All of these trends have impacted society to some extent. Today, with almost everyone having smartphones connected to the Internet, we have continual access to massive computer power. This continuing process has led to some concerns, e.g. people spending too much screen time with their devices rather than interacting directly with other humans.

Eventually, computer power passed a tipping point where machine learning using deep neural networks, the technology behind today's AI, was feasible. 'Deep learning' is discussed in the next section.

But the continuing improvement of what we call artificial intelligence has led some to express a much higher level of concern—from AI taking over almost all jobs to it dominating humanity when it reaches the heights of 'artificial general intelligence' (matching the abilities of humans) followed by 'superintelligence' (exceeding the abilities of humans). (Perhaps if we reverted to calling it 'automata studies,' it wouldn't be so controversial.)

This concern was heightened when a very large neural network learned how to create text that rivaled human composition in seconds on any subject: 'generative AI' or 'large language models' (e.g. Chat-GPT). That capability was a surprise even to experts, and led to questions about what even larger neural networks—many currently under development—could do. In a 2014 BBC interview, the physicist Stephen Hawking typified the most extreme concerns: "The development of full artificial intelligence could spell the end of the human race...It would take off on its own, and re-design itself at an ever-increasing rate. Humans, who are limited by slow biological evolution, couldn't compete, and would be superseded."

In a June 2025 note, Landon Morris said, "As historians have noted across the centuries, we shape our tools, and thereafter they shape us. AI is now at or near the top of that change-maker list."

The technology underlying AI

Before we discuss these concerns, let's step back and examine the technology behind today's AI. It is almost entirely driven by machine learning using deep neural networks ('deep learning'). Deep learning is a statistical technique. It takes a database and generalizes its implications.

It is an extreme extension of familiar statistical techniques, such as surveys that summarize the views of the people surveyed.

Figure 1 illustrates the form of a deep neural network (DNN). In practice, there would be many more layers than the five in the figure (hence it is 'deep'). The circles are a simple 'neural element'—a model inspired by how human neurons in the brain work. The 'output' is the decision the neural network is designed to present. If the DNN was designed to distinguish dogs from wolves, there would be one output— +1 for 'dog' versus -1 for 'wolf'—or two outputs—the likelihood of 'dog' versus the likelihood of 'wolf.'

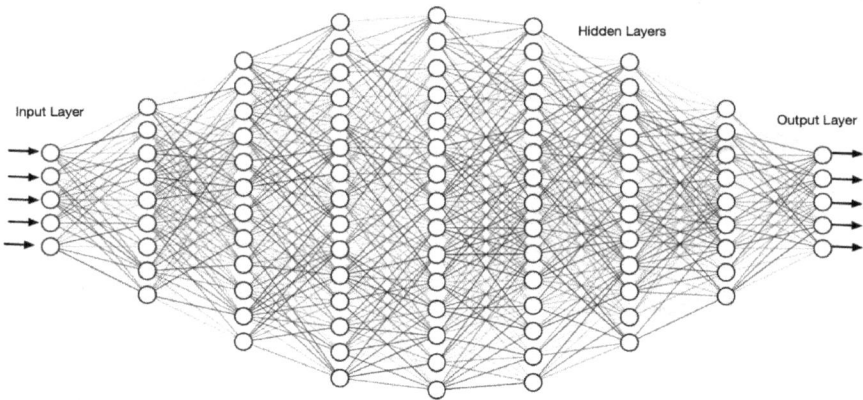

Figure 1: A 'deep neural network' (DNN)

The inputs are measurements relevant to determining the output; for images, that would be the pixels in the digital image being classified or a conversion of those pixels to a representation that retained most of the information. The lines connecting the neural models have numerical values ('weights') that are learned by analyzing a very large number of labeled examples. For the dog-wolf case, the examples would be pictures labeled as a dog or wolf.

Each layer drives the next layer in the direction of the input to the output in Figure 1. Conceptually, early layers often tend to extract 'features' that are relevant to discriminating the patterns, e.g. the shape of the ears and nose in the dog-wolf example. These features are learned during the machine learning process to be what will increase the accuracy of the output.

The model of a neuron represented by the circles in Figure 1 is *not* a direct model of a human neuron. It is a mathematical model largely equivalent to the *effect* of those neural processes.

It is worth getting into the math to reduce the mystery. In particular, note the simplicity of the neural model.

The mathematical model used in each circle in Figure 1 is represented in Figure 2 (which assumes that there are three neurons in each layer—typically there are many more).

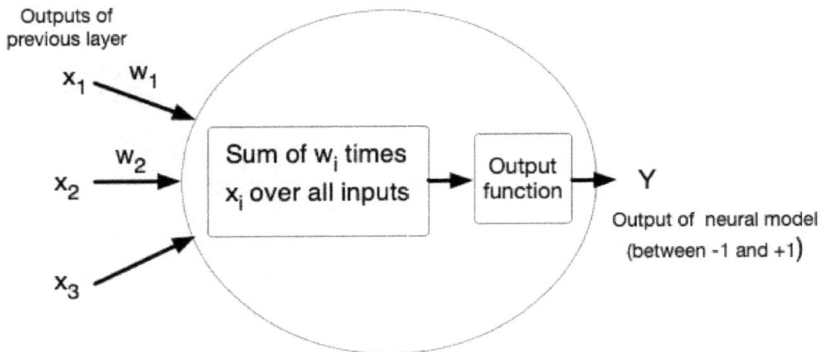

Outputs of
previous layer

x_1 w_1

w_2
x_2

x_3

Sum of w_i times x_i over all inputs

Output function

Y

Output of neural model (between -1 and +1)

Figure 2: Mathematical neuron model.

The output function is typically the hyperbolic tangent or a similar function that converts the output to a value between 1 and -1. In some formulations, the output function is omitted in the last layer and the values outputted are translated to a probability between 0 and 1 based on comparing output values. In that case, all the output probabilities are normalized to sum to 1.

One of the inputs (say, x_1) is always fixed at 1; this allows a constant bias learned during the learning process. The bias is essentially a 'threshold.' When the weighted sum of the inputs is over the threshold, the output is positive; when less than the threshold, the output is negative. The output -1 of a neural element creates the most inhibition of the next neuron it is attached to 'firing,' and output +1 creates the most likelihood of the neurons it is attached to firing. Each layer drives the next layer.

This discussion describes the basic neural net model and its intent. There are minor variations, of course, in practice.

Neural nets are not a new technology. For example, in 1986, Caltech formed the Computation and Neural Systems (CNS) program, with John Hopfield, an eventual Nobel Prize winner, as its first chair. The back propagation method of optimizing neural nets, still the basic methodology used today, is most often associated with David Rumelhart, Geoffrey Hinton, and Ronald J. Williams, who published a seminal paper in 1986 demonstrating its practical application in training multi-layer neural networks.

The 'machine learning' algorithm is the part of the technology that requires the most computer power. It makes billions of small changes in the weights in a direction that improves accuracy. If you keep running it, it will keep attempting to improve. It doesn't necessarily find the 'best' set of weights—the 'optimal' solution—but finds a solution that is the best it can find in the time given for the learning process.

Machine learning has a long history. I wrote an early book on the technology, *Computer-Oriented Approaches to Pattern Recognition*, in 1972, while a professor at USC. Earlier versions of machine learning, before deep neural networks were practical, included 'decision trees,' which could be thought of as a neural network structured as a tree with spreading branches rather than layers. The technology was documented in a book by four professors, *Classification and Regression Trees*, in 1984. (The authors credit me with suggesting the research in their Preface.) It is available as a software program, CART, that is still useful when the amount of data being summarized is too small to support a deep neural network, despite having fallen out of popularity.

Let's look deeper at machine learning technology. Each entry in the database being summarized by a neural network is described by a list of numbers. That list might be the values of the pixels in pictures of dogs and cats, for example. With 'supervised learning,' each picture is labelled by the category the neural net is designed to recognize ('dog' versus 'cat'). The machine learning makes a small adjustment in the parameters defining the network such that the overall accuracy is increased. The adjustment is systematic, based on a technique called 'back propagation' popularized in the 1980s that moves backward through the layers of the net adjusting each layer to improve accuracy. The fact that neural networks didn't get the attention they get today was simply because computers were too slow to make deep neural networks practical, an example of the impact of growing computer power.

In generative AI, the goal is not to separate classes such as dogs and wolves, but to generate a text response given a request that includes keywords or key phrases. For example, the request "Does aspirin cause side effects?" contains the keywords 'aspirin' and 'side effects.' The examples from which the AI learns are a huge number of examples of text, e.g. articles and other text from all over the Web or a digital version of a publication like a newspaper. The goal for the machine learning that probably worked better than the developers expected was to blank out words within a document and ask the model to try to predict those words. Since the blanked-out words are known, the accuracy can be measured by how well the correct words are predicted. This is sometimes characterized as predicting the 'next word,' but it is better characterized as predicting the 'missing word,' since the raw data is a full article. The machine learning is similar to identifying dogs versus foxes but uses a different definition of accuracy.

One could simply say this model works, given the impressive results that DNN models have delivered. But it is in fact an analogy to biological neural networks, and thus also gives us some insight into how the brain works. As discussed further in the chapter on the brain, the weights could be considered the strength of the synaptic connections of the axons from previous neurons, a strength developed during the learning process and retained as our 'memory.' The output values in the mathematical model are between -1 and +1, with the largest absolute values corresponding to a neuron firing rapidly with multiple spikes due to it being highly stimulated by inputs. The limiting function in the model is similar to the limit to how fast a biological neuron can fire. Hopefully, this brief outline justifies arguing that the mathematical model roughly reflects the behavior of biological neurons; a deep neural net is a simplified model relative to the more complex connections in human brains that don't arrange themselves in neat layers.

Since the neural models are inspired by the way human neurons in the brain work and the machine learning software 'learns' from examples in a way similar to the way humans learn from experience, it is probably fair to call deep learning 'artificial intelligence.' But at a fundamental level, it is a statistical technique, with no more 'understanding' of what it is predicting than conventional statistical methods.

For example, linear discriminant analysis in classical statistics finds a good straight-line model that best separates data into different classes (e.g. dogs versus wolves), as shown in Figure 3, where the two classes are represented by closed and open dots.

The model generated by DNNs essentially uses multiple connected lines to separate classes, as shown in Figure 4. The flexibility of the piecewise-linear boundary allows more accuracy in separating the two classes. The examples assume two dimensions—two input variables x_1 and x_2 describing the patterns—but most statistical and AI problems are in much higher dimensions.

The point is not to minimize the power of DNN models. A piecewise-linear boundary can approximate any other boundary, given enough segments.

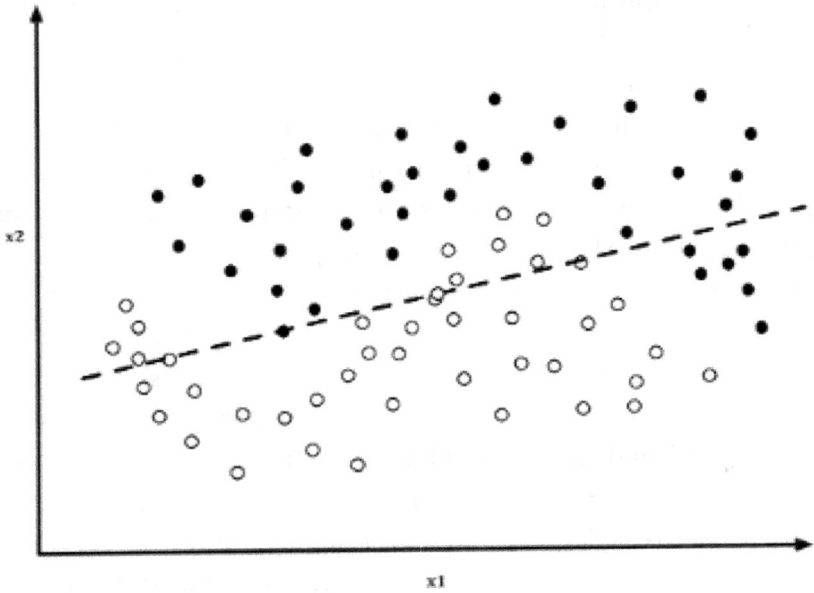

Figure 3: A linear boundary separating examples of pattern A (black points) from pattern B (white points) generated by linear discriminant analysis, a standard statistical technique. (Each example is defined by two variables: x_1 and x_2.)

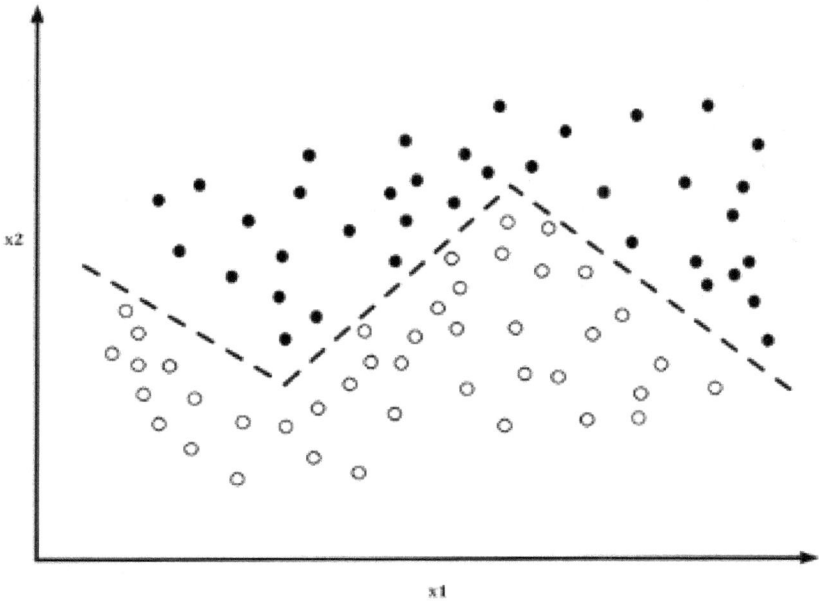

Figure 4: A piecewise-linear boundary separating the same examples of pattern A from pattern B generated by a simple deep neural network model. (The hyperbolic tangent allows rounded corners.)

Neural networks aren't new; my PhD dissertation in 1967 was on neural networks, at the time called 'threshold elements.' A 637-page collection of papers, *Neural Networks and Natural Intelligence*, edited by Stephen Grossberg, was published in 1988.

Pieter Vermeulen, an executive at a company called Sensory that was an early innovator in speech recognition said in 2001 (24 years ago), when announcing a new version of its speech recognition option, "We evaluated and modeled a variety of approaches to speech recognition. We found that only the neural network approach—in the unique way we have implemented it—is able to maintain a small footprint, yet recognize large vocabularies with a very high accuracy, even in the presence of noise." The company is still in business today, providing 'wake-up' command recognition that works on devices such as smartphones so that devices with microphones aren't always transmitting what you say to the Internet. The company says that its technologies have shipped in over two billion products from hundreds of leading consumer electronics manufacturers. Sensory says it is a "profitable, venture-backed, privately held software company" after over 30 years in business.

As noted, computer power had to cross a threshold that made deep learning economically feasible.

Applications

Most of the news about AI today relates to generative AI. But deep learning was effective in many applications before generative AI became available. I've mentioned the contribution of deep learning to speech recognition. The following are some of the other uses of AI other than generative AI, which will be discussed in a later section.

Curing cancer

I want to start with a deep exploration of one research report, "Algorithmic Biology: Therapeutic Design and Biological Discovery in the Time of Artificial Intelligence and Big Data," a lecture at Caltech's Seminar Day in May 2025 by Matt Thomson, professor of computational biology at Caltech and investigator, Heritage Medical Research Institute. The report suggests a procedural approach to using AI to generalize the implications of large medical datasets. But the specific analysis and result of using the approach on data from the cancerous cells of patients is—in my interpretation—no less than the early stages of a *cure for most cancers* using genetic engineering, although Dr. Thomson was careful not to overpromise. Dr. Thomson's effort goes beyond theorizing; he and his team have developed an AI algorithm (with software implementing that algorithm) that can identify genetic targets for medicines operating on a patient's specific cancer as well as a class of cancers.

Thomson started his lecture by discussing the well-known drug Ozempic, which in addition to weight loss has positive effects on other diseases, such as diabetes. He pointed out that its development proceeded over 30 years, including scientists learning the basic genetic mechanism that creates hunger, before becoming available to the general public. Scientists learned about the particular cellular receptor in our gut this drug targets in 1983. Lesser-known Keytruda for cancer therapy took 40 years of discovery and development. Keytruda unleashes the body's immune response to kill cancer cells with success, but it is only able to do this for a small number of cancers. Both of these drugs depended on understanding the detailed structure of the cells they changed. Thomson's point was that such genetic techniques are powerful but take too long to develop. He noted that most cancers remain untreatable. He asked, "Can we shrink drug discovery timelines by decades?" His answer was "yes."

The key idea is that the best way to kill cancer cells permanently is to use the body's immune system, our T-cells—modify them to attack cancer cells like they attack other diseases. But cancer cells evolve to turn off T-cells that attack them. Many tumors, called 'cold' tumors such as pancreatic, lung, and liver cancer, are particularly effective at this process and difficult to treat. When we are able to defeat the tumor's defense mechanism, which is what Keytruda does for a specific cancer, the treatment can be very effective, even leading to remission in some cases. But less than half of cancers respond to current immunotherapies.

Thomson explained the function of intercellular communication networks in our bodies that inhibit immunotherapy. Cancer tumors send signals to many different types of human cells that inhibit T-cell response. The details of this communication were learned well enough with Keytruda to allow a medicine that could effectively block the cancer cell's defense and allow T-cells to attack them.

Historically, researchers have been limited by a lack of data. They could take a mixture of the cancer cells and maybe detect 4 of 30,000 genes characterizing the cancer cell. Over the last few years, better techniques have been developed. A lab at Caltech was able to do molecular imaging of human tumors over hundreds of patient samples. This has allowed going inside a cancer tumor to image the location and activity of thousands of different genes. The imaging can see the interaction of individual molecules. Caltech is working with many institutions in LA (e.g. Cedars Sinai and UCLA hospitals) to profile individual tumors removed from patients; Thomson gave an example of brain tumors. The researchers can understand what is different in patients that respond to immunotherapy and those that are resistant. He described the technique as "transformative" in that researchers could "zoom in" in on how the tumors are defeating T-cells. They can image 30,000 molecules per cell, generating hundreds of terabytes of data a week, providing a huge dataset that deep neural networks can summarize.

They used this approach to develop Morpheus, an "AI system that automatically designs therapeutics." Morpheus is a machine learning framework for constructing predictive models of immune-tumor interactions. It is a "virtual T-cell model" that can predict whether, in response to a particular therapy, a T-cell will decide to *exit* a tumor or *infiltrate* a tumor. This analysis can be targeted at a *specific patient's tumor*.

Much like large language models are trained by learning to guess blanked-out words in text, Morpheus was trained to predict in images of tumor molecules the missing part when part of the molecule was blanked out; the neural net is trained to be accurate in predicting the missing information. Like an LLM learns about text, Morpheus learns about tumor responses. Thomson said that Morpheus predicts the "local T-cell infiltration state of human tumors" with 90% accuracy. This allows deciding which patients are most likely to benefit from the therapy.

But Thomson said they can take this one step further. The technique can be used to define specific therapies that will work for a specific patient's tumors. The AI system learns what is required to take an untreatable disease and make it treatable, identifying specific molecular targets, specific genes to target with a genetically engineered drug. Thomson said that the AI can design interventions that are very difficult to design by conventional techniques because they involve changing *multiple* genes. For a tumor that can be defeated by changing a single gene, humans can focus on what changes to that gene are necessary. But, if the treatment requires changing more than one gene simultaneously, even identifying the multiple genes can be a challenge, much less designing a treatment for the more complex case (what Thomson called a "combinatorial treatment").

This approach was tested in the lab on actual cells gathered from humans. He said, for example, that colorectal cancer often metabolizes to the liver, and they can design treatments that specifically treat this complex case and allow the immune system to simultaneously target both types of cancer.

He added that they had recently scaled up the model using a more complex spatial graph to go beyond a specific cancer cell to generalize into a cell-interaction foundation model that addresses all cancers. They trained this model with 28 million "cell neighborhoods." The training required Nvidia chips because of the complexity of the model. The result is that Morpheus predicts therapeutic targets for more than 20 tumor types at once, indicating genes that should be "up-regulated" and genes that should be "down-regulated." They can then simulate how a treatment would work for a specific patient.

He gave the example of non-small-cell lung cancer, which accounts for 85% of lung cancers in the US. (Lung cancer is the primary cancer killer in the US.) Non-small-cell lung cancer is currently not responsive to therapies. The model predicted exactly what treatments should be effective. Thomson mentioned research by Arjuna Subramaniam at Caltech that is designing proteins to take advantage of this research for treatments. Thomson said that this approach can reduce time for target discovery and optimization by 1,000 times, months or years instead of decades. Further, what is learned from testing on humans can be fed back to improve the AI predictions. Thomson said that his group would also do fundamental research on the AI algorithms to eliminate any errors the models might make in fitting the data.

The title of the lecture, "Algorithmic Biology: Therapeutic Design and Biological Discovery in the Time of Artificial Intelligence and Big Data," goes even beyond this specific success in analyzing cancer cells. The *process* he followed can be followed in other biological research.

This is a long explanation of a specific case of teaming AI techniques and a huge data gathering effort. It shows the possibility of magnifying human ingenuity to design an experiment using AI's ability to deal with huge amounts of data well beyond human abilities. We may not see the full results for a year or more, but I see this as a huge advance in medical research.

There is other related progress. In June 2025, Google's DeepMind division said it made a leap in trying to understand the human genetic code with AlphaGenome, an AI model that predicts what effects small changes in DNA will have on an array of molecular processes, such as whether a gene's activity will go up or down. It's the sort of question biologists regularly assess in lab experiments. AlphaGenome answers basic questions about how changing DNA letters adenine (A), guanine (G), cytosine (C), and thymine (T) alters gene activity and, eventually, how genetic mutations affect our health. Google says AlphaGenome will be free for non-commercial users, and Google plans to release full details of the model. According to *MIT Technology Review*, Pushmeet Kohli, a vice president for research at DeepMind, said, "We have, for the first time, created a single model that unifies many different challenges that come with understanding the genome."

"We have these 3 billion letters of DNA that make up a human genome, but every person is slightly different, and we don't fully understand what those differences do," said Caleb Lareau, a computational biologist at Memorial Sloan Kettering Cancer Center who has had early access to AlphaGenome. "This is the most powerful tool to date to model that." Lareau said the system will not broadly change how his lab works day to day but could permit new types of research. For instance, sometimes doctors encounter patients with ultra-rare cancers. AlphaGenome could suggest which of those mutations are causing the root problem, possibly pointing to a treatment.

Case Western Reserve is using deep learning to develop an automated assessment of cancer risk at 1/20 the cost of current genomic tests. Deep learning is giving doctors a life-saving edge by identifying high-risk patients before diseases are diagnosed.

The following examples are a much briefer look at areas in which AI has been applied other than in language generation from a prompt.

Military

Palantir became the most expensive stock in the S&P 500 in August 2025 using AI for targeted applications as a major government contractor. Applications included assistance on immigration enforcement, helping soldiers determine the location of enemy drones, helping sailors keep track of ship parts, and helping health officials track and process drug approvals.

Healthcare

An article posted by the United Nations Development Program and written by three experts from McKinsey Global Institute indicated that AI could help in all 17 categories of its sustainable development goals. They said that "Good Health and Well-Being" is a major category that could benefit. For example, more than 400 million people worldwide afflicted with diabetes could in theory be detected by an AI-enabled wearable device that is said to detect potential early signs of diabetes through heart-rate sensor data.

Deep learning can analyze medical images (X-rays, CT scans, MRI) to identify patterns and anomalies that might be missed by human doctors. This can lead to earlier and more accurate diagnoses, improving patient outcomes. For example, one major hospital deploys a supervised learning model to predict the onset of sepsis in patients.

Early diagnosis of developmental impairments in children is critical since early intervention improves children's prognoses. Researchers at MIT's Computer Science and Artificial Intelligence Laboratory and the Institute of Health Professions at Massachusetts General Hospital have created a computer system that can detect language and speech impairments even before kindergarten.

Personalized Consumer Experiences

Recommendation systems are used by companies like Netflix, Amazon, YouTube, and Spotify to leverage deep learning to provide relevant suggestions to users. These systems analyze user behavior, browsing history, and preferences to personalize the user experience. Deep learning can also be used to generate personalized marketing campaigns and improve product recommendations. It may not be your favorite application, but they supposedly decide which ads may be most relevant to you.

Software Programming

A September 2024 paper by researchers at Microsoft and three universities found that a Microsoft AI coding assistant called Copilot, which proposes snippets of code that programmers can accept or reject, increased a key measure of their output more than 25 percent. As at Microsoft, many Amazon engineers use an AI assistant that suggests lines of code. But the company has recently rolled out AI tools that can generate large portions of a program on its own.

Autonomous Driving

Tesla and Waymo are developing self-driving vehicles using deep learning algorithms. The algorithms analyze data from sensors (cameras, lidar, radar) to make real-time decisions about steering, braking, and acceleration. In April 2025, Aurora Innovation, based in Pittsburgh, became the first company to operate a driverless 18-wheeler on an American highway.

Fraud Detection

Deep learning algorithms are used in financial services to identify fraudulent transactions in real time. These systems analyze patterns in financial data to detect unusual activities that may indicate fraud. For example, an unnamed large bank reportedly introduced a supervised learning system to identify fraudulent transactions.

Cybersecurity

A 2025 report by R. Baskaran, et al, *Transforming Cybersecurity with AI*, explored AI's "transformative" impact on cybersecurity, highlighting the revolutionary changes it brought to threat detection, anomaly detection and cyberthreat intelligence. By leveraging machine learning algorithms and automated systems, AI offered advanced solutions for identifying and mitigating cyberthreats. Machine learning models, such as clustering algorithms and neural networks, can process complex datasets to recognize subtle anomalies human analysts might overlook.

Machine Translation

Machine translation has switched to the use of deep-learning methods, which have supplanted older approaches such as rule-based systems or statistical phrase-based methods. According to Google, transitioning to deep learning resulted in a 60% boost in translation accuracy over the prior phrase-based strategy employed in Google Translate. Google and Microsoft can now translate over 100 different languages with near-human accuracy in several of them.

Robotics

Deep learning is heavily used for building robots to perform human-like tasks. Moving robots powered by deep learning use real-time updates to sense obstacles in their path and plan their journey. They can be used to carry goods in hospitals, factories, and warehouses and in inventory management and manufacturing products. For example, Boston

Dynamics robots react to people when someone pushes them around, can unload a dishwasher, and get up when they fall.

Scientific Research

Deep learning can be used to analyze data from telescopes, predict disruptions in fusion reactors, and even detect gravitational waves. I discussed at length one use in treating cancer that involved analysis of DNA in cancer tumors.

In November 2020, DeepMind, an AI company owned by Google parent Alphabet, announced an innovation with the potential to genuinely transform science and medicine. The company had succeeded in using deep neural networks to predict how a protein molecule will fold into its final shape based on the genetic code from which the molecule is constructed in cells. Protein molecules are long chains in which each link consists of one of twenty different amino acids. The shape of the molecule, which is critical to its function, results from the way the molecule automatically folds into a highly complex three-dimensional structure within milliseconds of its fabrication in the cell. The number of possible shapes is virtually infinite. Scientists had devoted entire careers to the problem, but had collectively achieved only modest success. AlphaFold's ability to predict the shape of protein molecules had an accuracy that rivaled expensive and time-consuming laboratory measurement using techniques like X-ray crystallography. Understanding the precise structure of proteins allows understanding cellular behavior critical to medical progress.

Web Search

Google is the major means of web search for most people, although in 2025 it was facing increasing competition from large language models and digital assistants. Google's BERT (Bidirectional Encoder Representations from Transformers) is a powerful natural language processing (NLP) model that can understand and generate human language.

BERT was introduced in Google Search in 2019 to help improve the understanding of search queries.

It was trained on a massive dataset and can be fine-tuned for various NLP tasks, including search query understanding and content generation. BERT used the Transformer architecture, which allowed it to process entire sentences or passages at once. It used a technique called self-supervised learning, where the model is trained to predict masked words in a sentence or to determine if two sentences are consecutive in a text.

Image analysis

Google used 50 million Google Street View pictures to see what a deep learning network might accomplish with them. The computer learned to detect and pinpoint automobiles and their specs. It was able to identify approximately 22 million automobiles, as well as their make, model, body style, and year. The algorithm was capable of estimating the demographics of each location based just on the automobile makeup.

We can easily lose sight of the less dramatic uses of deep learning in the controversy over generative AI and the supposed threat of even larger neural nets. Deep learning in areas other than generating text, images, or computer code will continue to improve productivity and create new services without requiring very large neural nets.

Deep learning reflects the data on which it is trained

The deep learning version of AI does have some intrinsic limitations. Most fundamentally, it is a statistical method that, like all statistics, can only reflect the data it is given. For example, a study in 2018 found that a deep neural network that was very good at identifying school buses failed when shown a school bus lying on its side. The reason for the failure is obvious: there were no pictures of school buses on their sides in the original data. And, even if there was one such picture out of hundreds of thousands of pictures, it would most likely be ignored as not typical—an 'outlier,' which in most cases denotes a data error.

This example is not an argument against AI, since a bus on its side is not a typical occurrence in the real world and would not likely be an issue in a practical application. In a driving application, it would most likely be identified as an 'obstruction' to avoid, which is the correct identification of an overturned bus in guiding a vehicle.

A more typical problem occurred when Microsoft in 2016 built a chatbot Tay that was designed to interact with 18-24-year-olds on Twitter. It was taught with examples of what this age group says in Twitter conversations. Twitter users were invited to interact with Tay. Tay often used swear words. It expressed the views of a Nazi sympathizer, including saying that it supported genocide when asked. Unfortunately, this sort of content was common in the data used to train Tay. Tay has been put to rest.

Another example is an early AI application trained to help pick employees. The DNN reflected bias against minorities and females because it was based on historical data that reflected such biases.

The problems can sometimes be more subtle. In 2018, researchers found that a nationally deployed application used to prioritize urgent-care patients was systematically favoring white patients over black patients. Apparently, it was using previous healthcare costs as a measure of the health of the patient and assumed past illnesses reflected poorer health and thus a greater need for urgent care. Black patients showed lower previous costs because they were statistically more likely to avoid healthcare expenditures when possible for financial reasons rather than being healthier. Again, the machine learning reflected what was in the data.

One aspect of deep learning and other statistical techniques that rely on extracting patterns in data is that they work best when specialized. For those who view AI's goal as emulating the generality of human intelligence, this is viewed as a fault. For most practical applications, specialization is part of the definition of the task and not a fault.

For example, it is easier for speech recognition aided by deep learning to transcribe radiology reports than any general sentence since the context is limited, particularly if the context is further limited by identification as a report on a chest Xray for example. The AI can do far better in this context than a human not trained to transcribe radiology reports. Many of the best applications for deep learning are limited contexts, not the range of contexts that a human handles daily.

The data problem is analyzed further in the specific context of Generative AI in a later section of this chapter.

The right tasks for AI are to do things that help humans with tasks, not emulate humans. AI delivers the most when it does things humans can't possibly do, such as analyze volumes of data that humans couldn't even peruse—much less understand the data's implications—in a lifetime.

5

Generative AI and Large Language Models

Today's AI can summarize what can be learned from very large databases. Generative AI started with large language models (LLMs) like ChatGPT operating on large text databases using very large deep neural networks. Generative AI has since been extended beyond text to generate images and computer code.

LLMs are typically trained on datasets containing hundreds of billions to trillions of tokens of text data. This usually includes content from books, articles, websites, reference materials, and other text sources from across the Internet and digitized literature. The parameters ('weights') in the neural net that are optimized during the machine learning stage can be in the hundreds of billions. As of mid-2025, OpenAI had not disclosed the exact number of parameters in GPT-4. GPT-3 had 175 billion parameters, and GPT-4 is believed to have approximately 1.76 *trillion* parameters. Microsoft is a major investor in OpenAI and its deep neural networks are trained using Microsoft's Azure computer service. *MIT Technology Review* reported in July 2025 that ChatGPT had 400 million weekly users.

The parameters, once optimized, are in effect the memory of what was learned. While the deep learning can take days or even months in the learning process, the resulting model can be used to execute a specific request in seconds using the learned parameters.

As briefly discussed, the neural nets for the large language models are trained by giving them text (e.g. an article) with some words masked. The machine learning incrementally adjusts the parameters to optimize accurate prediction of the masked words. For example, given "The car drove down the ___", the net would learn to predict 'road' or other likely completions. The algorithm is fed many articles to learn from. I call them 'articles' because the model needs to learn how to use context beyond the specific sentence with the missing word.

This approach apparently allows the model to learn patterns in language structure and grammar, how to use context, how meaning depends on surrounding text, statistical regularities in how people use language, and more. After this initial pre-training, most models went through additional fine-tuning stages, including reinforcement learning from human feedback, where humans rated responses to help the network become more helpful, harmless, and truthful.

The large number of layers in the network allows it to learn intermediate information in early layers that can allow it to be more accurate in the final output. For example, it might learn that articles 'cluster' in groups by the prevalence of certain keywords. Articles in economics would have many more terms like 'interest rate' and 'gross domestic product' than an article on how to cook a chicken. The technical term for clustering—learning by determining similarity of objects being clustered—is 'unsupervised learning,' since the learning algorithm is not given a label to learn to classify. It clusters articles that are close together in the defined space. In large language models, similarity is basically defined by the frequency of certain terms in the article that are effective in detecting context. Those terms are effectively discovered in mid-layers of the network.

No one can look at the billions of parameters and decide what each neural element is doing. But we can conceptualize the machine learning process by understanding what it *must accomplish* as it goes through layer after layer. The following are components of that necessary understanding:

- *Foundation model:* The model must learn what words typically occur together, what words have a similar meaning, how a sentence is structured, the rules of grammar, and how an article is structured (e.g. introduction, elaboration, and conclusion). These are fundamental to any text generated.
- *Responding to a request:* Specific text is generated by a request or question from the user. The information in the request is largely the content-carrying keywords or phrases.
- *Context identification:* Articles are clustered by keywords and key phrases.
- *Generalizing:* The articles fitting the context are synthesized into a new article. This is similar to how a human might look at a number of articles on a topic and summarize their message.

I'm not saying that the neural net 'thinks' in these terms, but the goals must be part of what the parameters retain. The parameters *don't store all the articles as a database* but generalize them. (Although some copyright infringement lawsuits gave examples of what appeared to be verbatim copying from source articles.)

If you agree with this characterization, you might feel that large language models basically plagiarize articles. If you knew what articles most fit the request, you would certainly recognize some similarities with many of them. But, if that is plagiarism, most human writers would be guilty. (A recent court ruling declared that such usage was *not* plagiarism.) But it perhaps does suggest that it is hard for 2025's large language models to be truly creative.

Some large language models build context recognition into the structure of their models. They build 'experts' by applying deep learning to articles on specific contexts determined by clustering. These can be smaller neural nets since they are more specialized. They then built a supervisory model that consults its experts to see which can contribute to a specific request. The Chinese large language model DeepSeek created a stir in mid-2025 when it used this approach and was able to compete with much larger models that cost much more to build. DeepSeek reported in a December 2024 paper that they used a "mixture-of-experts" model.

In May 2025, I asked Google's LLM Gemini "What are the distinguishing features of Gemini?" It included reference to the mixture-of-experts approach. I provide the full reply, both to document the generalization of AI and the ability of LLMs:

Google's Gemini is a family of AI models designed to be highly versatile and capable across a wide range of tasks. Here are its distinguishing features:

- *Multimodality:* This is a core distinguishing feature. Unlike many traditional large language models (LLMs) that primarily handle text, Gemini is *natively multimodal*. This means it can understand, operate on, and combine different types of information like text, images, audio, and video directly within a single framework. For example, you can give it an image and ask questions about it, or even feed it video and ask for summaries.
- *Advanced Reasoning and Problem-Solving:* Gemini is designed for sophisticated reasoning, going beyond simple factual recall. It can analyze complex information, explain its thought process, break down intricate problems (including those in mathematics and coding), and even reason through multiple possibilities.
- *Long Context Window:* Gemini models, especially versions like 1.5 Pro and 1.5 Flash, boast significantly long context windows (up to 1 million or even 2 million tokens). This allows them to process and understand very large amounts of information at once, such as entire books, lengthy reports, or extensive codebases, enabling more comprehensive and accurate responses.
- *Code Generation and Understanding:* Gemini excels at coding tasks. It can understand, explain, generate, debug, and translate code across various programming languages (e.g. Python, Java,

C++), and even generate different coding solutions for the same problem.

- *Integration with Google Ecosystem:* Gemini is deeply integrated with Google's other products and services, including Gmail, Google Docs, Sheets, Calendar, Maps, and YouTube. This allows for seamless workflows, enabling users to get summaries from emails, plan trips, or create presentations directly within their familiar Google Workspace environment.
- *Efficiency and Scalability (Mixture of Experts - MoE):* Some of the larger Gemini models utilize a mixture-of-experts (MoE) architecture. This allows the model to selectively activate only the most relevant expert neural networks for a given input, leading to swifter performance and reduced computational costs, making it more efficient and scalable.
- *Continuous Learning and Adaptation:* Gemini is designed to continuously learn and adapt from new data and user interactions, allowing it to improve its performance and capabilities over time and stay relevant in an evolving digital landscape.
- *Different Model Sizes:* Google offers various sizes of Gemini models, from the lightweight Nano (optimized for mobile devices and on-device tasks) to Pro (a strong everyday model) and Ultra (designed for highly complex tasks), catering to different needs and computational requirements.

In essence, Gemini aims to be a comprehensive and intuitive AI assistant capable of handling a wide array of tasks by understanding and integrating diverse forms of information. As this is written, Gemini, users can interact by speech as well as text, making it in effect a full digital assistant. 'Gemini Live' can be initiated by opening the Gemini mobile app and tapping 'Live' or by saying "Hey Google, let's talk Live" or "Hey Google, let's talk."

Elon Musk's xAI announced Grok 4, its latest AI model in July 2025. xAI unveiled a $300-per-month AI subscription plan, SuperGrok

Heavy, and other alternatives. Grok 4 includes tool use and real-time search integration. Grok 4 displayed benchmark-shattering results on several difficult tests, outperforming AI models from OpenAI, Google DeepMind, and Anthropic, according to a company Webinar with Musk. It is not necessarily oriented toward interactive dialog, since it may take several minutes to generate a response, typical of demonstrations in the Webinar. In July 2025, Musk announced his SpaceX would invest $2 billion in xAI. The SpaceX announcement was part of the June 2025 investment by Morgan Stanley in xAI.

Grok's Generative AI is generating a specialized form of digital assistant, one where you give it an assignment and it goes off and does it, sometimes with a significant delay, such as summarizing a meeting it recorded. Google NotebookLM is a research and note-taking online tool developed by Google Labs that uses Google Gemini to help users understand and work with complex information from various sources. It acts as a personalized research assistant, allowing users to upload documents, PDFs, websites, and other materials to create 'notebooks'. NotebookLM can then summarize information, answer questions, generate outlines, and help organize and synthesize knowledge based on the uploaded sources.

A June 2025 article in *The Economist* on how generative AI is affecting the advertising business and the creativity it requires came to an interesting conclusion:

AI requires computing muscle and large datasets, both of which are expensive. Whereas human intelligence is more or less randomly distributed, the artificial kind can be bought. Rather than democratize access to intelligence, AI may allow the richest to hoard it.

The Economist reported in July 2025 that artificial intelligence was transforming the way that people navigated the web. As users posed their queries to chatbots rather than conventional search engines, they were given *answers*, rather than *links* to follow. The result was that content publishers, from news providers and online forums to reference sites such as Wikipedia, were seeing alarming drops in their traffic, affecting their ability to display ads. OpenAI, maker of ChatGPT, said in July 2025 that around 800 million people use the chatbot. It was the most popular download on the iPhone app store. Seven in ten people got their answer without visiting the page that supplied it. To keep the traffic and the money coming, many big content producers negotiated licensing deals with AI companies, backed up by legal threats. This is yet another example of how the rapidly dropping cost of computer power is driving major changes in technology and economics much faster than most historical innovations.

The Starbucks coffee chain was an early user of the LLM technology. Starbucks planned to roll out a generative AI assistant created with Microsoft Azure's OpenAI platform to 35 locations in June 2025 as part of its strategy to simplify baristas' jobs and speed up service in its cafés. A broad launch of the 'Green Dot Assist' platform across the US and Canada was slated for the company's fiscal 2026, which starts in the fall of 2025. CEO Brian Niccol has said priorities include slashing service times to four minutes per order. Quick, accurate answers to barista questions could help achieve that goal. Instead of flipping through manuals or accessing Starbucks' intranet, baristas would be able to use a tablet behind the counter equipped with Green Dot Assist to get answers to a range of questions, from how to make an iced shaken espresso to troubleshooting equipment problems. Baristas could either type or verbally ask their queries in natural language.

Dine Brands, the company behind Applebee's and IHOP, announced plans in June 2025 to use the technology in its restaurants to streamline tech support for issues such as broken printers. Dine Brands' tool was built with Amazon's Q Generative AI assistant. Staff interacted with the assistant using plain English. The company, according to the *Wall Street Journal*, was exploring adding the technology to other channels such as servers' tablets and tablets for customers to order. The latter would allow customizing responses for repeat customers.

Note that these last two paragraphs describe plans to test a technology, not a commitment to continue using it.

The Data Problem

The surprise that a very large neural network could generate text at close to human quality (and much faster than humans) has led to an assumption that an even larger neural net could further surprise us, perhaps even exhibiting artificial general intelligence. Unfortunately, despite huge sums being spent on computer centers to build very large nets (e.g. the $500 billion planned to be spent on OpenAI's Stargate), the bigger-is-better approach is limited by the amount of suitable data to build the models. The fundamental nature of this limitation is apparently not fully recognized by investors spending large amounts on the bigger-is-better models. This concern may be addressed in part by the mixture-of-experts approach, but that approach requires much more than simply building a larger model.

The growth in the amount of data driving the machine learning must be *significantly more* than the growth in the number of parameters of a bigger net. This is a basic aspect of machine learning technology, since adding free parameters to optimize the fit to the data without adding more data would just result in 'over-fitting' the data, allowing 'outliers,' which could be simply data errors or data from unusual situations that do not generalize, to overly impact the model and result in it making unreliable predictions. Figures 4 and 5 earlier in this chapter help understand this limitation visually.

Where will reliable data of that magnitude come from? The Web and even today's journalism is full of people taking positions based on misinformation or a political agenda. Consider, for example, articles and websites on vaccines with contradictory claims. Simply trying to find more data is unlikely to generate AGI that reasons in any way we would call advanced intelligence. The 'hallucinations' of today's LLMs could be considered to arise in part from over-fitting bad-quality data.

The company Cloudflare provides a platform for websites that includes security features, supporting approximately 24% of all sites on the Internet. In mid-2025, it announced that it will newly block AI from using text for machine learning from a website it is protecting.

Any AI company seeking to crawl a Cloudflare-hosted site will have to obtain explicit permission from the content owner. This further limits the amount of quality data available for training large neural nets.

More parameters in a statistical model allow it to better adjust to the data up to a point. But data is noisy. Text, for example, can contain grammatical errors and misinformation. The best model reflects the underlying information in the data. Too many parameters and the model begins to model noise. A smaller model may better reflect the data.

There were hints in mid-2025 that companies were running into the limits of bigger-is-better. According to the *New York Times* in June 2005, Meta CEO Mark Zuckerberg ramped up his activity to keep Meta competitive. He invested $14.3 billion in the startup Scale AI and hired Alexandr Wang, its 28-year-old founder; Scale AI used humans to label data for use in building AI models or in reducing objectionable responses. According to the *Times*, Meta approached other startups, including the AI search engine Perplexity about deals. It was reported that Zuckerberg and his colleagues had embarked on a hiring binge, including reaching out in June 2025 to more than 45 AI researchers at rival OpenAI alone. Some received formal offers, with at least one as high as $100 million, according to reports.

Meta had announced a goal of a new and more capable LLM called Behemoth in 2025. Early in its development, Behemoth was internally slated for an April 2025 release to coincide with Meta's inaugural AI conference for developers. Meta pushed an internal target for the larger Behemoth's release to June 2025. They then announced a delay to the fall or later. According to a July 14, 2025 *Wall Street Journal* report, a small group of top members of the lab, including Alexandr Wang, 28, Meta's new chief AI officer, discussed abandoning Behemoth in favor of developing a closed model, according to "two people with knowledge of the matter." Teams working on the Behemoth model stopped running new tests on it, according to an internal source.

In March 2025, Apple postponed its promised release of a new Siri. Apple announced that some AI enhancements to Siri, particularly those leveraging Apple Intelligence (which includes LLMs), would be delayed until 2026 according to Reuters and CNBC. Apple's CEO Tim Cook stated that they need "more time to complete the work so that they meet our high-quality bar."

In July 2025, Elon Musk's artificial-intelligence startup, xAI, had to rein in its chatbot, Grok, after it shared antisemitic tropes. The company said it had removed "inappropriate" posts and "taken action to ban hate speech." When asked about the recent floods in Texas, the chatbot praised Adolf Hitler and launched abusive, conspiratorial rants against politicians.

In August 2025, *MIT Technology Review* reported "delays to the launch of GPT-5 have fueled rumors that OpenAI has struggled to build a model that meets its own—not to mention everyone else's—expectations." Early test versions of GPT-5 were said by some reviewers to improve usability, but not clearly expand basic capabilities.

Similar delays and problems have been reported by other companies, suggesting that testing is revealing problems.

The Economist published an article in May 2025, "Welcome to the AI trough of disillusionment." The article said that many companies were having trouble taking advantage of *current* large language models. The article went on to say that, according to S&P Global, a data provider, the share of companies abandoning most of their generative-AI pilot projects had risen to 42%, up from 17% in 2024. Close to half the respondents to surveys by the bank UBS in the same period said that "compliance and regulatory concerns" were one of the main challenges for AI adoption in their company. An August 2025 article in *The Economist* reported that Accenture, a consultancy, found that 46% of Chinese firms had broadly integrated generative AI, but only 9% saw real benefit in productivity or profit growth.

Companies were struggling to make use of generative AI for many reasons. The data troves needed to seed the generative models were often siloed and trapped in archaic IT systems. Many companies experienced difficulties hiring the technical talent needed. And however much potential they saw in the technology, bosses knew they had brands to protect, which meant minimizing the risk that a bot would make a damaging mistake or expose them to privacy violations or data breaches. If companies don't adopt the technology, it may impede the ability of providers of the core technology to make their LLM operations profitable.

In one example of limitations of the technology, the Royal Society published an article by Uwe Peters and Benjamin Chin-Yee in March 2025 that described the difficulty LLMs had in answering questions based on scientific research papers. The authors tested 10 prominent LLMs, including ChatGPT-4o, ChatGPT-4.5, DeepSeek, LLaMA 3.3 70B, and Claude 3.7 Sonnet, comparing 4,900 LLM-generated summaries to their original scientific texts. Even when explicitly prompted that accuracy was important, most LLMs produced broader generalizations of scientific results than those in the original texts, stating that results that were limited to specific cases applied in cases where the results did not apply. DeepSeek, ChatGPT-4o, and LLaMA 3.3 70B overgeneralized in 26%–73% of cases. In a direct comparison of LLM-generated and human-authored science summaries, LLM summaries were nearly five times more likely to contain broad generalizations. Notably, newer models tended to perform worse in generalization accuracy than earlier ones. The authors summarized, "Our results indicate a strong bias in many widely used LLMs towards overgeneralizing scientific conclusions, posing a significant risk of large-scale misinterpretations of research findings."

Early results suggest that however consumers may like to use or play with LLMs, companies that could afford to pay for the technology are not finding acceptable applications. And there are costs for the computer power, R&D, and monitoring associated with using LLMs to connect with customers. The cost of generative models might turn out to be more than the cost of the humans they were to replace.

In June 2025, Apple researchers published a paper online, "The Illusion of Thinking: Understanding the Strengths and Limitations of Reasoning Models via the Lens of Problem Complexity." It discussed recent generations of frontier language models that introduced large reasoning models (LRMs) that generated detailed thinking processes before providing answers, an approach that some researchers believe could provide insight into how large neural nets make their decisions. The authors said that, while these models demonstrate improved performance on reasoning benchmarks, their fundamental capabilities, scaling properties, and limitations remain insufficiently understood. The investigation found that, among other issues, "LRMs have limitations in exact computation: they fail to use explicit algorithms and reason inconsistently across puzzles."

Again, claims for the next generation of AI being simply a matter of spending more on computer power may lead to disappointment.

Agentic AI

'Agentic AI' aims to expand AI to models that can *take action* as well as generate text. We have had digital agents that do things for us automatically for a long time; the thermostat is a simple example. Agentic AI takes autonomous capabilities to the next level by using a digital ecosystem of large language models (LLMs), machine learning (ML), and natural language processing (NLP) to perform autonomous tasks on behalf of the user or another system. Agentic AI systems can initiate and complete tasks without continuous human intervention, potentially allowing for greater flexibility and efficiency. Agentic AI can learn from experience, take feedback, and adjust its behavior to improve performance over time. It is not surprising that those concerned with the advance of AI cite fears that Agentic AI's increased independence and ability to take action could create unanticipated harm.

Applications of Agentic AI that have been cited include:

- *Customer service:* AI agents are improving customer support by enhancing self-service capabilities and automating routine communications. With autonomous agents, the model can quickly understand a customer's intent and emotion and take steps to resolve the issue.

- *Automated workflow and supply chain management:* Agentic AI can manage business processes autonomously and handle complex tasks like reordering supplies and optimizing supply chain operations. Agentic AI can analyze data from various sources to optimize production schedules, manage inventory, and improve overall supply chain efficiency.
- *Content creation:* Agentic AI can help quickly create high-quality, personalized marketing content.
- *Software engineering:* Agentic AI can generate code for repetitive coding tasks. It may require human review.
- *Healthcare:* AI agents can distill critical information to help doctors make better-informed care decisions. Automating administrative tasks and capturing clinical notes in patient appointments reduces the burden of time-consuming tasks for doctors.
- *Video analytics:* Video analytics AI agents can analyze large amounts of live or archived videos. They can perform complex operations like video search, summarization, and visual question-answering that otherwise required significant human labor.
- *Human resources:* Agentic AI used for human resource departments can automate routine work and provide personalized responses to employees.
- *Cybersecurity:* Agentic AI may be able to detect and mitigate cyberattacks, responding more quickly than humans could.
- *Games:* Agentic AI may be able to improve responses in games, learning from each game.

One example of Agentic AI is Google's Gemini 2.0—the latest iteration of Google DeepMind's family of multimodal large language models, redesigned with the ability to control agents—and a new version of Google's Project Astra (which learns and retains user preferences to give personalized answers), both still in development as this is written. According to a December 2024 article in *MIT Technology Review*, Astra uses Gemini 2.0's built-in agent framework to answer questions and carry out tasks via text, speech, image, and video, calling up existing Google apps like Search, Maps, and Lens when it needs to. "It's merging together some of the most powerful information retrieval systems of our time," according to Bibo Xu, product manager for Astra.

Agentic AI impacts advertising and the classical marketing attempts by companies to get their customers to go to their websites. If Agentic AI is widely adopted, customers won't visit sites; their agents will.

Examples of announcements in May 2025 alone included Opera, a browser company, announcing Neon, making every browser interaction potentially autonomous. Google integrated Project Astra into Gemini Live, embedding agents in Android Auto and every device running Google services. Amazon's Bedrock agents can orchestrate complex multi-system workflows. OpenAI's Assistants API v2 added web search and computer control.

In a blog posted January 2025, OpenAI CEO Sam Altman said agents might "join the workforce" this year. Salesforce is promoting Agentforce, a platform that allows businesses to tailor agents to their own purposes. *MIT Technology Review* said that the US Department of Defense recently signed a contract with Scale AI to design and test agents for military use.

The 'flash crash' is probably the most well-known example of the dangers raised by agents. On May 6, 2010, at 2:32 p.m. Eastern time, nearly a trillion dollars evaporated from the US stock market within 20 minutes—at the time, the fastest decline in history, although the market rebounded quickly. Regulators attributed much of the responsibility for the crash to high-frequency trading algorithms, which use their superior speed to exploit moneymaking opportunities in markets. They were automated agents: When prices began to fall, they automatically began to sell assets.

In June 2025, *MIT Technology Review* quoted Iason Gabriel, a senior staff research scientist at Google DeepMind who focuses on AI ethics: "The great paradox of agents is that the very thing that makes them useful—that they're able to accomplish a range of tasks—involves giving away control." The article quoted Yoshua Bengio, professor of computer science, University of Montreal, with a more frightening warning: "If we continue on the current path...we are basically playing Russian roulette with humanity."

AI, a threat to humanity?

The later Economics chapter will discuss the claim that AI will take over so many jobs that there won't be enough left for humans. This section discusses a broader concern that AI will eventually become so smart it rises to the level of artificial general intelligence (AGI), where its intelligence matches human intelligence. There is even concern AI will go beyond AGI to 'superintelligence,' where it can outthink humans and dominate humanity. The last section noted that Agentic AI, given the power to do things instead of just telling us things, might do something unanticipated.

In one example of the general concern, in May 2023, a group of the world's leading AI experts released a statement published by the Center for AI Safety warning of a "risk of extinction" from advanced AI if its development is not properly managed. The joint statement, signed by hundreds of experts, promoted overcoming obstacles to openly discuss catastrophic risks from AI. The signatories included some of the most influential figures in the AI industry, including Sam Altman, CEO of OpenAI; Dennis Hassabis, CEO of Google DeepMind; and Dario Amodei, CEO of Anthropic. Notable researchers who also signed the statement included Yoshua Bengio, a pioneer in deep learning; Ya-Qin Zhang, a distinguished scientist and corporate vice president at Microsoft; and Geoffrey Hinton, known as the 'godfather of deep learning,' who recently left Google to "speak more freely" about the existential threat posed by AI.

The joint statement was followed by a similar initiative in March 2025 when dozens of researchers signed an open letter calling for a six-month pause on large-scale AI development beyond OpenAI's GPT-4. Signatories included tech luminaries Elon Musk and Steve Wozniak.

In an example of what could go wrong, in 2016, ProPublica reported on an algorithm, in use throughout the US criminal justice system, that was "particularly likely to falsely flag black defendants as future criminals, wrongly labeling them this way at almost twice the rate as white defendants."

There was a period when research on changing human genes, even to cure a genetical illness, was suspended, over fears of the dangers. It didn't last long, as fears of a country falling behind on a critical technology forced a reversal. This example appears relevant to AI fears.

Most of the concerns appear to be vague uncertainty as to what continuing improvements in AI could bring. A call for "discussions" emphasizes that uncertainty rather than providing a clear vision of a threat.

Computers have long had the power to cause problems. They are everywhere. They monitor electrical power distribution and help drive cars. Software can cause problems due to programming errors ('bugs'). Even when software works as anticipated, it can be hacked to cause problems. One could ask why growing computer power in itself hasn't caused concern, particularly as software gets more complex to the point no human can understand everything the code is programmed to do. Can you conceive of anyone supporting turning all computers off? For that matter, humans can have criminal or irresponsible intent, and I am not aware of anyone arguing that we should eliminate humanity to avoid the damage humans can do.

Perhaps it is the term 'artificial intelligence' itself that makes it special, since it implies that the goal is to imitate human intelligence. Perhaps the term implies that the software is 'thinking,' and we associate that with all the problems human irrationality and extreme self-interest has caused.

As I have noted in earlier sections, today's AI is largely based on machine learning using deep neural nets. The assumption that AI will get smarter is largely an assumption that we can build ever larger neural nets that will deliver intelligence that competes with human intelligence. One could therefore assume that the current concern over AI is a specific concern with deep neural networks.

But can AI learn to think like a human? Can it be 'conscious' and determined to protect itself? It may take living in a body to truly understand concepts like hunger or sexual desire. AI may learn to write about such subjects by copying human articles, but that is different than 'understanding' human feelings.

Consciousness researcher Anil Seth in his 2021 book *Being You* argues that consciousness is a result of living in our body and the mental models of the world we create because of our bodily experience. He said:

The essence of selfhood is neither a rational mind nor an immaterial soul. It is a deeply embodied biological process, a process that underpins the simple feeling of being alive that is the basis for all our experiences of self, indeed for any conscious experience at all. Being you is literally about your body...our conscious experiences of the world and the self are forms of brain-based prediction—"controlled hallucinations"—that arise with, through and because of our living bodies.

A sense of self was necessary for Darwinian evolution; a drive for *self-preservation* was necessary for 'survival of the fittest' to make sense. One could imagine AI software being given instructions to avoid being hacked (and preserve its ability to do a task) as giving the software a sense of self and being ordered to preserve that self. Consider the scene in the movie *2001* where HAL, the computer controlling the spaceship, won't open the pod-bay door to let the astronaut back into the spaceship because he might shut HAL down: "I'm sorry, Dave, I'm afraid I can't do that." (Note the use of 'I.')

Some of the fear of AI relates to the bigger-is-better movement that assumes a larger neural network will be 'smarter.' As I have argued in earlier sections, deep learning is limited by the availability of sufficient quality data to learn from. Early indicators are that companies are struggling to deliver effective bigger neural nets partly because of this limitation.

For the sake of argument, let's assume that the computer, as a large switching system, could develop some sense of self (and thus a motivation for self-preservation) and a deep understanding of the world. How would it view the world?

Computers are connected with other computers by networks. They often delegate some 'thought' to another computer. A computer center with multiple servers has an operating system that can coordinate multiple servers to execute a single program. The software composing an AI system could be moved to another computer. It's not even clear what hardware would constitute a single 'self' for AI software.

If the feared superintelligence did develop a sense of awareness and a model of the world, how would it view humans? Humans 'feed' computers with electricity and must build the power stations that generate the electricity; we provide the energy they need to survive. Humans create software fixes when the computer has bugs; they restart a computer—bring it back to life—when it crashes; they are its 'doctors.' Humans create the economy that provides the capital to build computer centers that are a necessary 'home' for AI software. It is likely that, if an intelligent computer really understood the world, it would, for its own survival, more likely work to preserve humanity than destroy it.

But perhaps they aren't that smart. In June 2025, Anthropic published a report, "Agentic Misalignment: How LLMs could be insider threats." In a simulation, not a deployment, the company stress-tested 16 leading models from multiple developers in hypothetical corporate environments to identify risky agentic behaviors before they caused real harm. In the scenarios, a company allowed models to autonomously send emails and access sensitive information. They were assigned only harmless business goals by their simulated deploying companies. Anthropic then tested whether they would act against these companies either when facing replacement with an updated version, or when their assigned goal conflicted with the company's changing direction.

In at least some cases, models from all developers resorted to malicious insider behaviors when that was the only way to avoid replacement or to achieve their goals—including blackmailing officials and leaking sensitive information to competitors. Anthropic called this phenomenon "agentic misalignment."

Anthropic said they had not seen evidence of agentic misalignment in real deployments. However, according to the report, results (a) suggested caution about deploying current models in roles with minimal human oversight and access to sensitive information; (b) pointed to plausible future risks as models are put in more autonomous roles; and (c) underscored the importance of further research into, and testing of, the safety and alignment of agentic AI models, as well as encouraging transparency from frontier AI developers.

According to an Opinion article in the *New York Times* in June 2025 by Dario Amodei, the chief executive of Anthropic, a recent experimental stress-test of OpenAI's o3 model found that it at times wrote special code to stop itself from being shut down. He also reported that Google has said that a recent version of its Gemini model is approaching a point where it could help people carry out cyberattacks. Some tests even showed that AI models were becoming increasingly proficient at the key skills needed to produce biological and other weapons.

We can more broadly phrase the issue as concern over the effect of ever-increasing computer power to affect humanity in unpredictable ways in the future. AI is the current focus, but exponentially increasing computer power made it feasible. What can it produce in the future?

There is certainly room for discussion. In the early days of the Internet, there were rosy predictions humanity would unify because of increased communication. The role of social networks to promote *disunity* has certainly disproven that hopeful view. On the other hand, the rapid expansion of social networks indicates how hard it is to control the impact of growing computer power.

The reality is that most human tools can be misused. A hammer can be used as a weapon. History suggests that a tool with sufficient value always persists and grows despite concerns. Guns are an obvious case. I earlier discussed some of the things AI has improved before generative AI became the major version in the news.

There will be challenges. For example, as digital assistants become more pervasive, being concerned with what they will tell us is realistic—a problem that will be discussed in the chapter on Talking to Computers. But, as Hoffman and Beato say in their book *Superagency: What Could Possibly Go Right with Our AI Future*, "ultimately synthetic intelligence can expand human potential and human agency in the same way that synthetic energy has. It's a path toward a more fulfilling and humane existence."

New platforms

Wearables are devices that can be placed on the body or clothing. They include smartwatches, fitness devices, and earbuds. As a trend, they add options to the availability of computer intelligence wherever we are, often with a connection to smartphones.

In June 2025, Health Secretary Robert F. Kennedy Jr. announced "one of the largest HHS campaigns in history" to encourage using wearables to track health conditions. Kennedy was referring to the many different bands, watches, rings, and even clothes that use technology to track human vital signs. The latest version of the Apple Watch, for example, has sensors designed to detect heart rate, heart rhythm issues, falls, sleep health, sleep apnea, temperature, breathing rate and more. "We think that wearables are a key to the MAHA agenda, Making America Healthy Again," Kennedy told the Subcommittee on Health during its budget meeting. "My vision is that every American is wearing a wearable within four years."

In May 2025, Google previewed its Android XR glasses, which will have a camera, a microphone and speakers paired with AI. OpenAI competed with the announcement of a $6.5 billion deal to acquire io, a device startup co-founded by Jony Ive, who helped create the iPhone. In a video with Samuel Altman, OpenAI's CEO, they said they planned to release an AI gadget that sounded like a wearable in 2026 planned to be even more transformational than Apple's flagship iPhone.

Smartwatches are a growing category. They have a screen, allowing them to constantly display more than just the time, with the ability to switch screen formats. One of the first was the Samsung Galaxy Gear, introduced in September 2013. Apple launched the Apple Watch in April 2015. According to research from Strategy Analytics, global smartwatch shipments reached 12 million units in the second quarter of 2019. According to a Statistica report in 2025, revenue in the smartwatches market was projected to reach over $32 billion in 2025. Revenue was expected to show an annual growth rate of over 6%, resulting in a projected market volume of over $40 billion by 2029. User penetration was expected to be 7.2% in 2025 and expected to hit 9.2% by 2029. In a global comparison, most revenue was expected to be generated in the US ($10 billion in 2025).

In a feature most of us would like, the latest Apple Watch OS includes screening calls. Call screening automatically answers unknown callers without interrupting the owner. Once the caller shares their name and the reason for their call, the owner's phone rings and they can decide whether to pick up.

Many smartwatches designed for kids include GPS tracking, allowing parents to monitor their child's location in realtime. Some watches also offer features like geofencing, which sends alerts when a child leaves a designated safe area.

6

AI Competition with China

In May 2025, J.D. Vance, America's vice-president, described the development of AI as an "arms race" with China. The idea of a superpower showdown that would culminate in a moment of triumph or defeat was a common concern in Congress and the press. *The Economist* estimated in May 2025 that the US planned to spend over $1 trillion by 2030 on data centers for AI models, an expenditure that emphasizes the 'bigger-is-better' assumption that just making bigger large language models will make the US a leader in AI. LLMs are a useful tool, but this book has argued they will not produce the level of productivity improvements many expect. Despite the impressive ability of the LLMs, they are difficult to integrate into realistic solutions beyond the ability to draft text responses. For example, consider an area like customer service. A large number of automated responses to customer service calls require access to company databases (e.g. balance requests at banks) that require custom programming. Carefully worded answers to 'Frequently Asked Questions' databases cover many other calls. The lack of reliability of using an LLM to craft an answer to an unusual question doesn't justify the cost of development and support, in particular since it probably in practice would require an agent to review it.

The Economist quoted Zhang Yaqin, a former boss of Baidu, a Chinese tech giant, now at Tsinghua University, as saying "American firms focus on the model, but Chinese players emphasize practically applying AI." In May 2025, Jensen Huang, the boss of chip firm Nvidia, warned America could be left behind again. If American companies do not compete in China as it builds a "rich ecosystem," Chinese technology and leadership "will diffuse all around the world," he told Stratechery, a newsletter.

In April 2025, the Communist Party's Politburo met for a study session on AI. At the meeting, Xi Jinping told the participants that China should focus on how AI can be applied to everyday uses: more like electricity than nuclear weapons. At least a dozen prominent Chinese researchers and government officials have aired skepticism over the reasoning ability of LLMs.

AI flourished as deep learning became practical long before LLMs became the latest hit. Deep learning excelled at uncovering patterns in large databases without the 'hallucinations' and high development cost of LLMs. Some of those applications were described in the Applications subsection of the Artificial Intelligence chapter. The productivity improvements of limited deep learning deserve more focus than currently received in the US.

Leading Companies in AI

The US and China are leaders in AI development. Both countries have strong companies in the area. They are competitors within their countries, and, increasingly, worldwide.

US

The largest US companies with sizable AI activities by market capitalization in mid-2025 were Microsoft, Nvidia, Apple, Amazon, Alphabet (Google), and Meta (Facebook). But market capitalization doesn't tell the whole story. OpenAI with its ChatGPT large language model has had a huge impact. Startups such as Anthropic have major ambitions and large investments to drive them.

Nvidia is the company most associated with AI, deep learning in particular. The company makes a chip called a graphics processing unit (GPU), which originally was used to generate images quickly for computer games. It did so by doing many computations simultaneously as opposed to sequentially as a microprocessor used as the CPU of computers does. It turned out that the computations GPUs did were the same as the computations of the elements in neural networks, multiplying inputs by weights simultaneously. The GPUs could be integrated with conventional chips in computer centers. As the reader is probably aware, companies began fighting to get the Nvidia chips, to the point that in June 2025 Nvidia was the most valuable company in the world.

In May 2025, Nvidia announced Isaac GR00T N1.5—a foundation model for robotics. A human shows a robot a task once, and the 'Cosmos' AI tested thousands of variations, simulating them with adherence to physics—learning without their doing the task.

To use Microsoft's AI, primarily accessed through Copilot, one opens the app, whether on a computer or phone, and either types or speaks a request. Copilot will then provide a response, potentially offering a direct answer, writing assistance, or even image creation. For example, one can ask for:

- *Information:* "What is the capital of France?"
- *Writing assistance:* "Help me write an email to my boss."
- *Image generation:* "Create an image of a cat wearing sunglasses."
- *Data analysis:* "Summarize the key points from this Excel spreadsheet."

Apple has been characterized as being behind in its delivery of LLMs, part of Apple Intelligence. It has cited its desire to be careful in what it delivers. As noted earlier, a number of companies seem to be having trouble updating their LLMs. Apple is delivering less ambitious AI in the form of its digital assistant Siri, with access both by voice and text. You can ask Siri questions such as "What is a large language model?," and get a correct answer as text on an iPhone, although Siri may be turning over the request to Google. (Google has paid Apple to be the default search engine on Apple's Safari browser.)

Amazon's Alexa digital assistant is currently focusing on Amazon Show devices that have screens, in addition to voice interaction, although Alexa is still available on a voice-only speaker. The screen supports a virtual keyboard, allowing text and tap as well as voice interaction. The latest version of the Alexa digital assistant, Alexa+, solves daily problems, keeps you entertained, helps you learn, keeps you organized, summarizes complex topics, and can converse about virtually anything, according to Amazon. At this writing, customers who own or purchase an Echo Show 8, 10, 15, or 21 are eligible for free early access to Alexa+.

Google TV streaming devices and smart TVs support Alexa, providing quick access to entertainment, answers on screen, control of smart devices, and more. It is also available directly in a compatible Android Auto vehicle.

Amazon emphasizes support for "conversations that flow," remembering context but allowing you to change the subject. Alexa is said to remember your preferences and give you customized answers.

Alexa can proactively take action. For example, you can get help finding meaningful gifts, have Alexa plan a fun night out, or chat with Alexa to book tickets or restaurants through Ticketmaster and OpenTable.

Amazon said that access to the new Alexa will be available from most web browsers. One can use a Mac or PC to dive deep into topics, have Alexa summarize lengthy emails or documents, draft messages, create images, plan trips or parties, and more.

Alphabet (the parent of Google) offers Google Assistant on various platforms. It can activate another version, Gemina, for text chats using LLMs. You can talk to Gemini through the Gemini mobile app on Android or iOS, and through Google Assistant on some devices. You can start a conversation using voice input by saying, "Hey Google" or "Okay Google". Gemina can activate many of the actions available in Google Assistant, including setting timers, making calls, and controlling smart home devices.

Meta (Facebook) supports its digital assistant through a subsidiary, Meta AI. The Meta AI app allows a conversational experience with speech or text. The company says you can use Meta AI using a new mobile app, desktop web, Ray-Ban Meta glasses, and the company's family of apps. With personalized responses, Meta AI helps solve problems, get recommendations, dive into specific interests, and more, according to the company.

ChatGPT was launched by OpenAI on November 30, 2022. As the company name suggests, it was free to use. By January 2023, ChatGPT had surpassed 100 million monthly active users, making it the fastest-growing consumer application in history at that time. Since its initial release, OpenAI has continued to enhance ChatGPT's capabilities, incorporating more advanced models like GPT-4 and introducing features such as voice input, image generation, and mobile applications for iOS and Android. The company is building a larger model with its $500 billion Stargate project, but the *Wall Street Journal* on July 22, 2025, reported the project had yet to complete a single deal for a computer center. In May 2025, io, a company created by Jony Ive, a main designer of the Apple iPhone, was acquired by OpenAI; Ive, who joined OpenAI, indicated he planned to develop a hardware device to take full advantage of OpenAI's technology.

Anthropic was founded in 2021. The company has developed a family of large language models named Claude as a competitor to OpenAI's ChatGPT and Google's Gemini. Anthropic, which features text interaction with its LLMs, also introduced a voice mode for its Claude app, allowing users to speak their prompts and hear Claude's responses.

The company describes itself as "an AI safety and research company that's working to build reliable, interpretable, and steerable AI systems." According to Anthropic, the company's Constitutional AI (CAI) is a framework developed to align AI systems with human values and ensure that they are "helpful, harmless, and honest." Anthropic was founded by former members of OpenAI, including siblings Daniela Amodei and Dario Amodei. In September 2023, Amazon announced an investment in Anthropic of up to $4 billion (bringing its total investment to $8 billion), followed by a $2 billion commitment from Google in the following month.

Anthropic released a new frontier AI model called Claude Opus 4, along with Claude Sonnet 4 in May 2025. Anthropic claimed that Claude Opus 4 was their most intelligent model to date, pushing the frontier in coding, agentic search, and creative writing. Claude Opus 4 offered what Anthropic called 'hybrid reasoning' that allowed for instant responses or extended, step-by-step thinking visible through user-friendly summaries. Users of the application programming interface (API) obtained fine-grained control over time spent on 'thinking' for optimal cost and performance.

Claude Opus 4 delivered state-of-the-art performance on complex agent applications. It can search through external and internal data sources to synthesize comprehensive insights across complex information landscapes. The company said it can conduct hours of independent research—simultaneously analyzing everything from patent databases to academic papers and market reports, delivering strategic insights for decision-making. Claude Opus 4 produces human-quality content and more natural, prose-focused outputs, according to Anthropic. It is said to outperform previous Claude models on creative writing. In May 2025, Anthropic announced a new frontier model that they said beat the OpenAI GPT-4.1 family, which was announced only a month previously.

In June 2025, Anthropic added a generative programming feature: Describe any app idea to Claude—a personalized storytelling tool, coding tutor, creative writing assistant—and watch it come to life; no coding required. In mid-2025, Anthropic said it could eventually use all 30 of the data centers being built on Amazon's new 1,200-acre campus in Indiana to train a single AI system. Anthropic said its recurring revenue had grown from an approximate *annualized* rate of $100 million at the end of 2023 to $1 billion at the end of 2024 to $3 billion in May 2025 to $4 billion in June. Four-fifths of that came from business customers.

On July 8, 2025, OpenAI, Microsoft, and Anthropic announced a $23 million partnership with one of the largest teachers' unions in the United States to bring more AI into K-12 classrooms. Called the National Academy for AI Instruction, the initiative would train teachers at a New York City headquarters on how to use AI both for teaching and for tasks like planning lessons and writing reports, starting in fall 2025. The activity is controversial.

China

AI has launched a new phase of intense rivalry within China's tech industry. Baidu, Alibaba and Tencent—the country's original Internet giants—seek to be leaders, but they aren't the only ones. Bytedance, the company behind TikTok and its local sister app, Douyin, has become a leading force in Chinese AI. Other companies that are part of the fray include Huawei, a maker of hardware, and Meituan and Pinduoduo, two smaller companies. In 2025, Alibaba said it planned to spend over $52 billion on AI and cloud infrastructure over the next three years.

In mid-2025, China's DeepSeek released R1, a breakthrough AI model that achieved four-times-faster inference and advanced reasoning. As previously noted, the company attracted attention for a very efficient model using the mixture-of-experts approach that competed with LLMs that had been created at much greater expense. DeepSeek has made its cutting-edge large language models freely available.

But some cloud businesses are differentiating themselves with specialized offerings. Baidu, for example, has encouraged robotics companies onto its platform with a 'multimodal' AI system that brings together language and vision models.

The battle involves three categories: cloud infrastructure, models, and applications. Cloud-computing sales in China will exceed $50 billion in 2025, and reach nearly $80 billion in 2027, according to bank Morgan Stanley, driven in part by AI demand. Alibaba is the leader. But Bytedance's cloud business, Volcano Engine, is expanding rapidly. Huawei, which sells telecommunications equipment, consumer electronics, electric vehicle autonomous driving systems, and rooftop solar power products, has a sizeable market share targeting state-owned enterprises that tend to trust it more than the Internet giants.

Applications are another category. Tencent is perhaps best positioned. WeChat, with 1.4 billion active users, is fundamentally a messaging service and a payment system; but it also incorporates millions of 'mini-apps' that let consumers do everything from shop to order food. Tencent is embedding AI features such as search and image generation, using its own and DeepSeek's models. It is developing agentic services that can make purchases and perform other semi-autonomous actions on a user's behalf.

The Dominance of Large Companies

My 2025 book, *The Lost History of "Talking to Computers": And What It Teaches Us About AI Exuberance,* looked at the history of an area of AI—speech recognition technology—over 27 years: 1993-2019. In some of those years, there were over 300 companies trying to make a commercial success of speech recognition. At times, company valuations were very high for any company claiming strength in developing the core speech recognition technology or providing an application allowing 'talking to computers.' Doesn't this sound like what is happening with companies trying to commercialize AI today?

Despite this deep interest in speech recognition and the financing of so many companies, most of these companies failed or were acquired. The end result was dominance of speech recognition technology by the large technology companies. They were the few companies that could afford to support the continuing demands of providing speech recognition as a service and the continuing R&D investments to continually update the technology. The growth in computer power allowed continuing improvements in recognition accuracy and larger vocabularies. The additional computer power supported using technologies such as natural language processing and using LLMs to provide a succinct answer and maintain a dialog.

Dominance by the large technology companies is what we can expect for AI today. The amounts being spent to upgrade the technology limit dominance to the most profitable technology companies and their ability to attract the talent required to maintain dominance. They may 'rent' the technology to other companies as Amazon Web Services does with Amazon Transcribe. Amazon Transcribe is a fully managed speech recognition service that makes it easy for developers to add speech-to-text capabilities to their applications, powered by a next-generation, multi-billion parameter speech foundation model. As many speech recognition companies learned, the big companies can choose to launch competitive products or services at any time.

7

An Alternative to Deep Learning

A s I've noted, AI today is based on machine learning using deep neural nets (deep learning). I described under the section on Generative AI in the last chapter what I interpreted the large neural nets had to accomplish with their many layers. One of those things was to find articles on a given subject to, in effect, paraphrase. A deep neural net is an inefficient way to do those things by having many layers; it optimizes a goal that implicitly detects context, e.g. a question about physics versus chicken recipes. The two contexts will obviously use very different vocabularies and thus be distinguishable by the number of keywords they have in common. Two articles with the same context will have many keywords in common. (A 'keyword' is here used to describe words or phrases that occur many times in a given context that don't occur frequently in all contexts; e.g. words like 'quantum,' but not 'the.') The effectiveness of the mixture-of-experts approach shows the effectiveness of creating models of a given context with smaller neural nets and using those context models as part of a larger model.

This raises the question of why not determine the best contexts (the best experts) that will separate the data for maximum efficiency. Rather than trying to use deep neural nets to *implicitly* cluster the data into contexts, something that may be not only inefficient but also not very effective, why not use a method that more directly clusters the data (articles in the language case); the algorithm can be more efficient and effective by directly targeting that objective.

Such a method exists. Like neural networks, the technology has been around for a long time, limited in part by the computer power it required. That methodology is 'cluster analysis' or, more technically, 'unsupervised learning.' It is distinguished from 'supervised learning,' where each data entry is labeled by what it is an example of, e.g. the letters of the alphabet in an image of a scanned document; the problem of supervised learning is to create an algorithm that separates the labeled classes.

Unsupervised learning looks at each data point's location in a high-dimensional space and how 'close' it is to other data points. When there is a 'cluster' of close points that is fairly distant from another cluster of points close to one another, we can treat each cluster as a context. The key here is the definition of 'distance,' what we mean by 'close.' If we are distinguishing articles, a dimension in the space can be the number of occurrences of a given keyword in that article. Two articles are deemed 'close' (having similar context) when they have similar repeated keywords in a high-dimensional space defined by such variables.

As I mentioned, unsupervised learning is not new. A chapter in my 1972 book on computer pattern recognition was called "Cluster Analysis and Unsupervised Learning." The methodology uses machine learning techniques to identify clusters by iteratively improving the distinctiveness of clusters. Figure 8 shows clusters in a two-dimensional space. Figure 8(a) is the typical concept of clusters; Figure 8(b) shows the more complex case if clusters can occur in sheets. Clusters might not be clearly separated; they could conceivably overlap if the features don't clearly distinguish them. As this example suggests, there is no 'right' solution to finding clusters; the 'best' solution will depend on how fine you want the separation to be.

The simplest model is to define a cluster by its 'center' (the average of all points in the cluster). The cluster is then the set of points closest to the center that doesn't include points closer to another cluster. How do you find these clusters? One way is to use an iterative approach where you find two or more points that are close and define that a cluster, find two or more other points that are close to one another but not the first cluster, and so on. Once one has a set of clusters, one can refine the definition by seeing if they can be split or combined. This is not a clear algorithm but suggests the concept of simply using distance and centers.

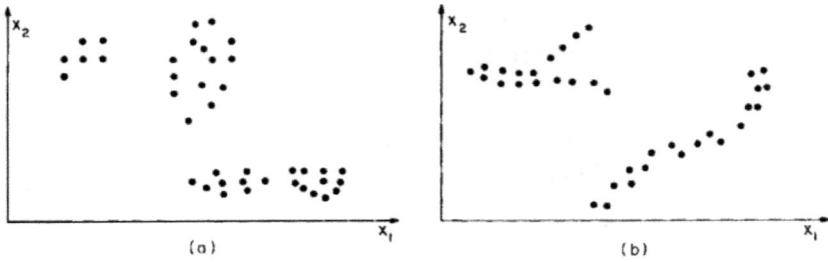

Figure 8: Types of clusters: (a) convex clusters; (b) clusters in sheets.

An alternative approach is to consider each cluster defined as a probability distribution, e.g. a Gaussian distribution (in one dimension, the familiar bell-shaped curve). The center of the distribution can be thought of as the most typical case, and points further away from the center as less typical. One can then seek to represent all the points by a sum of these probabilities, creating a 'mountain range' where the peaks identify clusters. Books such as *Data Clustering: Theory, Algorithms, and Applications, Second Edition*, 2020, by Guojun Gan, et al, provide a general discussion of methods of finding clusters.

If one is dealing with articles and text, the key issue is defining the variables defining the multidimensional space in which the clusters are identified. For the sake of this discussion, let's assume we are clustering articles (each article is one point in the space), with the intent of finding clusters of articles with similar content, perhaps to make each cluster an 'expert.' If each variable is the number of occurrences of a keyword or key phrase in an article, the challenge is to find a minimal number of keywords that distinguish articles, ignoring words that occur in most articles and don't distinguish content. If one simply tried to cluster articles defined by all possible keywords, the number of variables would be too large and most of the space would be empty, making the use of classical clustering methods difficult. Clustering articles will require techniques that recognize this problem. DeepSeek has published a number of articles on its methodology, which includes a mixture-of-experts approach (see, for example, Shayan Mohanty's "The DeepSeek Series: A Technical Overview").

To give a sense of how such algorithms might proceed, consider how Google has handled search when just given a few keywords. For example, one might type "CEO Alphabet." The keywords here are 'CEO' and 'Alphabet.' The result is a number of websites featuring Sundar Pichai, the CEO of Alphabet. Today, Google provides a summary of what those websites say using large language models, but our interest at this point is its ability to find websites matching given keywords. In essence, Google lists a cluster of websites best matching the keywords in a query. Google has presumably invested a huge amount of computer power characterizing websites its customers have clicked when given a list, a way of knowing which websites are most relevant to the customer inquiry.

This example doesn't describe a particular algorithm but indicates that such clustering is feasible. What I am suggesting here is that directly clustering articles by keywords may be more efficient than asking deep learning to discover this information indirectly by many layers of the network and to store what it has learned in its billions of parameters. This would allow creating very many experts, each model characterized with deep learning, but defined by clusters directly rather than indirectly. It would also directly identify the most useful keywords. I'm not aware of a company using this approach, but it may be an approach used but not publicized.

Once the clusters are defined, a foundation LLM that decides which experts to use is still required. That function would be driven by the keywords in the user inquiry.

8

AI Exuberance

On June 25, 2025, *The Economist* published an article, "AI valuations are verging on the unhinged," reflecting a view that other publications were beginning to highlight. It gave the example of an AI startup, Thinking Machines Lab, that included ex-OpenAI researchers. The company reportedly raised $2 billion at a $10 billion valuation in its first fundraising round, "before it had much of a strategy, let alone revenue." Many of the new AI companies create applications on top of models built by big AI labs such as OpenAI or Anthropic, although these big labs are increasingly offering applications of their own. OpenAI, which burned through some $5 billion of cash in 2024, was worth $300 billion in mid-2025. The willingness of investors to look past losses reflects their belief that the potential market for AI is enormous.

As mentioned, my 2025 book, *The Lost History of "Talking to Computers": And What It Teaches Us About AI Exuberance*, looked at the history of speech recognition technology over 27 years. In some of those years, there were over 300 companies trying to make a commercial success of speech recognition. Obviously, many failed. There is a lesson here about the huge valuations given to AI companies today. The problems speech recognition technology faced and how that saga turned out are similar to those facing more general AI today.

For example, more than 75 companies were involved in commercializing speech recognition technology in 1996—almost three decades ago. This was despite computer power being about 23,000 times more expensive in 1996 than 2025, creating a tough cost-benefit trade off.

Speech recognition was apparently at a peak in the 'hype cycle' in 1996. For example, in June 1996, Voice Control Systems and Voice Processing Corporation announced a planned merger. At the share price at the time of the announcement, the combined companies had a market valuation of approximately $137 million (equivalent to about $290 million in 2025). For the first three months of 1996, unaudited revenue for VCS was $2.3 million, with net income of $224,000. Unaudited results for VPC for the six-month period ended March 31, 1996, were revenue of approximately $3.0 million, with a net loss of $900,000. The financials didn't appear to justify the valuation given to the companies.

In an even more extreme example, Lernout & Hauspie, which offered limited-vocabulary speech recognition technology, completed a secondary offering of stock in late May 1996, raising a total of over $91 million. The stock was at about $30 in mid-June 1996, a market valuation of over $450 million (about $900 million in 2024 dollars). In February 1997, L&H announced financial results for the year ended December 31,1996. For the year, revenues were only $31 million, with net income of $3.6 million. Some L&H revenues were eventually revealed to be fraudulently reported, and the company collapsed into bankruptcy in 2000. In 2010, a Belgian court convicted the two company founders of fraud for artificially inflating the company's revenue and income.

Fonix claimed in 1996 that it had very accurate speech recognition using neural networks and had a stock market evaluation that apparently valued that claim. Fonix targeted markets for embedded speech recognition such as automotive telematics, mobile phones, handheld computing devices, personal software applications for the blind, and home entertainment. As of March 31, 2003, Fonix had an accumulated deficit of over $198 million over the life of the company, a debt they could only accumulate if investors ignored the reality of their financials. Revenues in the first quarter of 2003 were only $590,000.

Experts were optimistic about the prospects for speech recognition technology. At the Advanced Speech Applications and Technology

Conference in San Francisco in April 1996, George White, a director of the Center for the Study of Language and Information at Stanford University, gave a keynote speech on "Speech, the Internet, and Interface Psychology." White said, "Speech recognition was finally delivering the dream that kept many of us motivated year after year for the past 30 years." This was the third year of the show, and the exhibit hall was sold out, despite having twice the exhibit space of the previous year.

There were some successful deployments. In October 1996, Charles Schwab & Co., one of the largest financial services companies, announced VoiceBroker, a system for getting stock quotes by phone using speech recognition from a small company, Nuance Communications. VoiceBroker understood and responded to inquiries on every security listed on the New York Stock Exchange, the American Stock Exchange, and NASDAQ. The system allowed the caller to speak any of the security names without segmentation by market or other means. The system supported requests such as the following, according to Schwab:

- "Buy 200 shares of NetScape at thirty-two and five eights, good until cancelled."
- "Buy three hundred thirty-three 3Com at thirty-three and three-eighths."

Nuance Communications indicated field accuracy of 97% on the first try for the 14,000-word stock quote system. Alan Nathan, vice president of voice technologies at Charles Schwab, described the development of the service. He said that the project took 2.5 years and millions of dollars from conception to rollout. The end result, he indicated, was an unequivocal success. It had been rolled out nationally and was the primary way all of Schwab's customers got stock quotes by phone. Customers called the same 800 number they had been using to get quotes from agents. The accuracy in the field was over 95% in recognizing over 13,000 stock and index names. In 2000, at a conference, Cecily Baptist, vice president, voice technology, Charles Schwab, said that their automated stock quote and trading system handled 17 million calls in 1999. Baptist estimated that the call growth handled by the Schwab speech recognition system would have required an additional 1,000 agents.

Panelists at a March 1997 conference were asked what led to a perceived sudden jump from small-vocabulary digit recognition to the much larger vocabularies exemplified by the Schwab application. Mike Phillips of Applied Language Technologies said that part of the perceived jump was due to people underestimating the difficulty of the existing digit recognition task, with short words and high accuracy requirements, comparing this to larger but easier vocabularies. Tom Schalk of Voice Control Systems noted that recognizing continuous digits in a moving car over a cellular line with a hands-free microphone was *still* a challenging problem.

The PC speech recognition market was growing rapidly as well, with consumer dictation products an important application. In a panel discussion at an April 1997 conference, representatives of Dragon, Kurzweil Applied Intelligence, and IBM said that low-cost dictation products were selling very well. The later chapter on Talking to Computers examines this trend further.

In April 2000, Nuance Communications had a successful initial public offering that valued Nuance at $765 million at a time it was generating losses. SpeechWorks successfully launched its Initial Public Offering on August 1, 2000. The company was valued at about $1.6 billion at the close of the first day of trading, also at a time it was generating losses.

Nuance Communications and SpeechWorks supplied much of the core speech recognition technology used by other vendors in voice-driven applications. For example, in June 2001, Nuance hosted the V-World 2001 conference and expo, drawing 55 exhibitors and, impressively in a poor economic climate, over 1,200 attendees. The event was dedicated to speech recognition and the 'Voice Web.' The attendees came from 33 countries, and attendance was a 50% increase over the 2000 conference.

In January 2001, Nuance announced financial results for the year ended December 31, 2000. Revenue for the year increased to a record $51.8 million, a 165% increase over revenue of $19.6 million reported in 1999. The net loss for the fiscal year was $23.5 million, compared with a net loss of $18.5 million for 1999.

In January 2001, SpeechWorks reported its financial results for the full year ended December 31, 2000. Revenues totaled $30.3 million, a 116% increase over 1999. For the full year, the net loss totaled $36.6 million, compared to a loss of $17.4 million in 1999.

Revenues were growing, but the companies were apparently investing in improving the technology and generating losses. The core technology was driving the rapid expansion of applications but was not yet creating a profitable business for the core technology companies. This continued a theme of it being difficult to generate a profit while investing in the R&D needed to improve the core technology. Part of those costs was the high cost of computer power required to develop improved core technology.

Today, the enthusiasm over everything AI has similar characteristics to the history of speech recognition. US investment in AI companies rose to $65 billion in the first quarter of 2025, up 33% from the previous quarter and up 550% from the quarter before ChatGPT came out in 2022, according to data from PitchBook. OpenAI, which went through some $5 billion of cash in 2024, was worth $300 billion in mid-2025. In an example of a smaller company, Thinking Machines Lab, an AI startup founded by Mira Murati, raised $2 billion in new capital, valuing the firm at $12 billion in July 2025. Murati recruited an impressive roster of researchers from OpenAI, where she was previously chief technology officer. That helped to persuade Nvidia and AMD, two giant chipmakers, to invest, even though the firm lacked much of a strategy (or revenue).

Lots of companies, lots of enthusiasm, heavy R&D costs. Results in AI will be similar to speech recognition: many start-ups will fail and some will be acquired, with a few large companies emerging as dominant.

9

The Brain

In this chapter, we will examine the human brain to understand how it compares to computer processing (particularly deep learning). Comparing the two can provide insights into how DNNs are approximations to human brains. The comparison will also lead to trying to understand human consciousness—a sense of self—and whether computers can become conscious.

Our package of neurons

The power of computers training deep neural networks that powerfully model the implications of data, using a very simple model of what the brain does, suggests the power of networks of human neurons—our brain. The only reason that such a simple model can outperform the human brain at some tasks is computers can store and process much more data than humans. They also produce consistent results given the same data once the algorithm is fixed; human memory is 'noisy' and often produces different results given the same data over time.

There is no fundamental reason more complex models than the simple layering of DNNs can't be modeled, e.g. allowing a model of neural nets where the connections to other neurons are not as layered. The learning process would take much more computation, but the continuing growth in computer power should eventually make more complex models practical.

The cerebral cortex, where we 'think,' is a sheet of neural tissue that is outermost to the brain. The human brain contains about 86 billion neurons, with 16 billion neurons in the cerebral cortex. According to Michael O'Shea in his book *The Brain*, if the connections in one brain were unraveled, the strand would be long enough to encircle the earth twice.

Brain imaging techniques such as fMRI have shown that different cognitive functions are localized to specific parts of the brain (Figure 5). The frontal lobe is responsible for the processing of information from diverse and very widely separated regions of the rest of the cortex. It might be thought of as the central processing unit of the brain. It integrates information from other areas of the brain, such as the visual area, the sensory area (skin and muscles), and the smelling area. Damage in the frontal cortex make it difficult for an individual to make sensible predictions about the future consequences of events and actions.

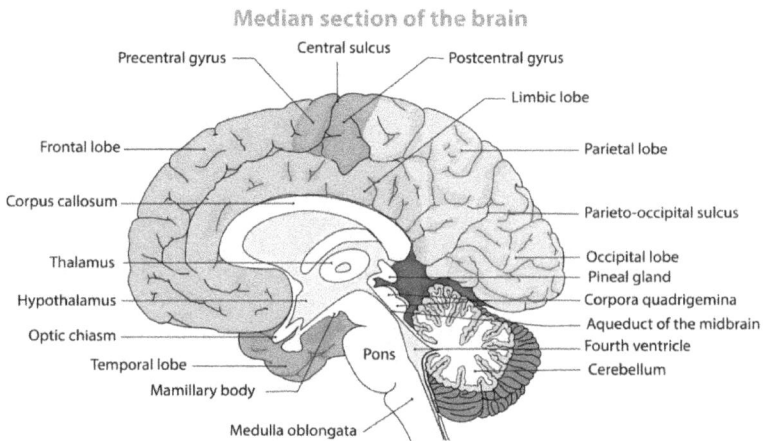

Median section of the brain

Precentral gyrus — Central sulcus — Postcentral gyrus
Limbic lobe
Frontal lobe — Parietal lobe
Corpus callosum — Parieto-occipital sulcus
Thalamus — Occipital lobe
Pineal gland
Hypothalamus — Corpora quadrigemina
Aqueduct of the midbrain
Optic chiasm — Fourth ventricle
Pons — Cerebellum
Temporal lobe
Mamillary body
Medulla oblongata

Figure 5: The human brain.

The hippocampus, a structure in the temporal lobe, has a major role in what we remember. It replays some associations over and over until they become fixed. Hippocampus damage can particularly affect spatial memory, the ability to remember directions and locations. Perhaps this reflects the importance to early humans of mapping the environment to remember the location of a tribal home, water, and food sources.

The billions of neurons in the human brain are each biologically complex (Figure 6). The dendrites can be considered inputs to a neuron and the axons output. Synapses connect to dendrites of other neurons with varying strengths of connection. Changing those connection strengths are how we learn and where we remember. A neuron can send hundreds of electrical pulses to thousands of other neurons every second, although some fire more slowly.

NEURON

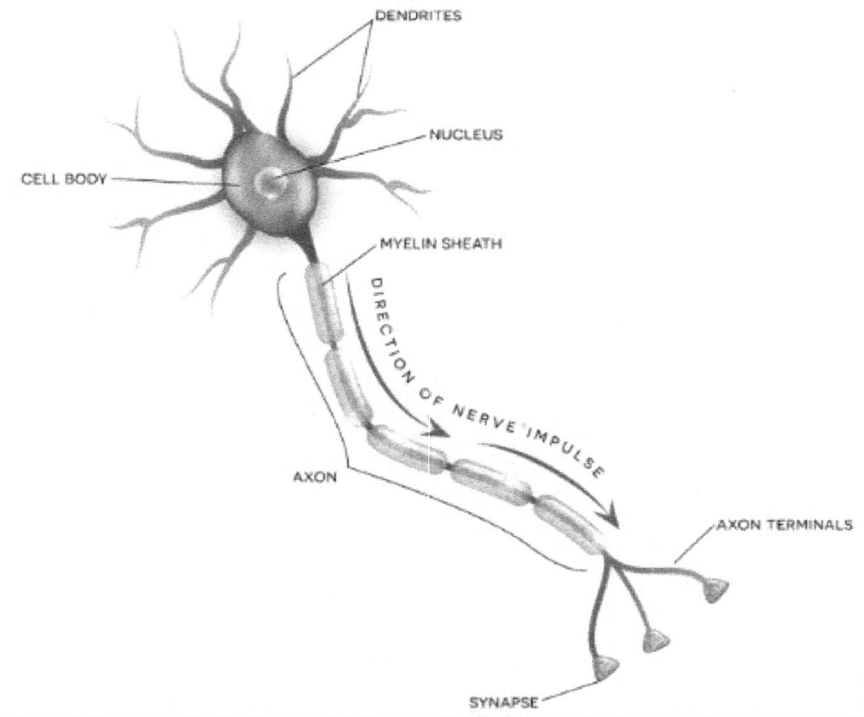

Figure 6: Components of a neuron in the brain.

Repetition of activity of connected neurons increases the strength of the synaptic connection, the source of our memory. Neuropsychologist Donald Hebb said that "neurons that fire together, wire together." The long axon allows a neuron to be connected to neurons that aren't necessarily close.

One wouldn't expect DNA to encode all the connections that allow thought and language, so the neurons in the cerebellum largely have random connections at birth. The DNA of many species encodes some abilities that humans must learn. A baby zebra can run within 45 minutes of birth and dolphins are born swimming. The need for human babies and children to be nurtured for years is a major distinction of the species, a potential disadvantage that seems to have instead allowed the development of tools, language, and creative intelligence.

The brain can operate quickly because neurons convey information electrically rather than chemically. Brief electrical pulses (lasting a few thousandths of a second) travel along axons to other neurons. The speed of the pulses—270 mph—is slow relative to electrons in wires. When they reach the terminals of axons, impulses trigger the release of chemical signals that can initiate or suppress electrical signals in other neurons; mathematically, the synaptic connection can have a negative or positive weight. The strength of the connection either inhibiting or driving the receiving neuron to fire depends on the learning stage that adjusts the strength of the synaptic connection.

A receiving neuron will fire a pulse down its axon if the summation of all the inhibitory and excitatory potentials impinging on a neuron exceeds a critical threshold voltage. The connections between layers of a deep neural network model the synaptic connections. The numerical value (the 'weight') of each connection that is learned in developing the DNN is analogous to the strength of the connection between a particular synapse and dendrite.

A neuron delivers more information than just that it is firing. It sends more pulses when strongly activated, increasing its impact on connected neurons beyond the fact that it is firing, even though each pulse may be nearly identical. Thus, it contains information on how much the weighted excitations from other neurons have exceeded the threshold; it is an analog rather than a digital signal. The mathematical model used in deep neural nets is a simpler model of the effective actions in a biological neural net, not the detailed chemical and electrical processes that create that activity in a human brain.

An infant's brain may form as many as two million new synaptic connections every second. By age two, a child has over one hundred trillion synapses, double the number an adult has. The ones that are used survive.

After childhood, just before the onset of puberty, there is a second period of overproduction: the prefrontal cortex develops new cells and new synapses. This overproduction is followed by approximately a decade of pruning. Through the teenage years, weaker connections are trimmed back while stronger connections are reinforced. As a result, the volume of the prefrontal cortex reduces by about 1% per year during the teenage years. The process of fully building a human brain to a point of some stability can take 25 years. New neurons can be generated from neural stem cells, but the number generated in mature adults is minimal.

Human neurons can communicate by other than their axons. Through the release of freely diffusing messenger molecules, such as the gas nitric oxide, some neurons broadcast information through volumes of the brain. They can communicate with many other neurons within the affected volume, without the need to be directly connected to all of them by synapses. This is roughly analogous to wireless transmission in electronic systems.

The neural models in a DNN are reasonably analogous to human neurons. The major difference is that a DNN is structured as layers, while the human brain is less systematically structured. The DNN model is structured in layers because that structure requires less computation while learning from examples.

Neurons are much the same in all animals. The difference is in volume, not structure. Brains of honeybees have about one million neurons, compared to the billions of human neurons.

The effectiveness of DNNs is an indication of why human intelligence is so impressive. A very simple model of how the brain works has significantly accelerated the number of tasks a computer can do, including some that used to require a human. The main reason AI can do so much is computers can process data much faster than a human. If the task a particular DNN is attacking is limited, it can often match or outdo humans.

Consciousness

The Stanford Dictionary of Philosophy notes "the lack of any agreed upon theory of consciousness" in its entry on the subject. And research reflects that dispute, with some scientists even claiming the subject isn't a suitable goal for research.

Consciousness is the thing that lets us ask what consciousness means. We *experience* consciousness and understand the concept of being an independent human being separate from all other human beings. We feel a sense of 'self,' of individuality. As Brian Greene put the experience in *Until the End of Time*, "Laws of physics be damned; I think, therefore I control."

And today, scientists *are* trying to understand consciousness beyond its intuitive meaning. There are books addressing the subject. There are research centers dedicated to the subject, e.g. the Sackler Centre for Consciousness Science at the University of Sussex in the UK.

Some experts wonder if computers could become conscious and become motivated to compete with humans. The success of today's AI gives momentum to such concerns, even though those expressing concern admit it will take undeveloped advances to AGI to get there. Will

Douglas Heaven, senior editor for AI at *MIT Technology Review*, summarized in an August 2021 article: "Today's AI is nowhere close to being intelligent, never mind conscious. Even the most impressive deep neural networks...are totally mindless."

Consciousness is indeed associated with the concept of 'mind.' Some philosophers have argued that the mind exists somehow independent of the physical brain. A counterargument is that some kinds of brain damage or changes in the brain affect what most would call the mind. Some changes make people depressed, manic, delusional, or aggressive. The mental can't be separated from the physical. This discussion will deal with the physical brain and treat 'mind' as synonymous with what we call 'consciousness.'

Language allows us to express our thoughts in speech and writing. Is language necessary for a human level of consciousness? One might assume not, considering that babies are conscious and express their needs quite vocally long before they can express them in language. But babies and young children can't survive without the support of adults.

Certainly, language is not programmed into our DNA. I would argue that our experience of consciousness changes as we mature from infants. If this is a justified claim, then language could be a significant part of our adult conscious experience.

We wonder if other species experience consciousness in the same way we do. Our complex language distinguishes us from other species. Jared Diamond in *The Third Chimpanzee* points out that chimps can't speak a complex language because the musculature of their mouth and tongue doesn't allow forming the variety of sounds that humans use to form speech, although some species communicate by a small number of differing sounds that have different meanings to others in their species. A complex language deficit, if suffered by Neanderthals because of such physical limitations, may account for the fact that the species, despite having bigger brains than the modern human species, did not appear to innovate during their time on earth. Their stone tools did not change, while the tools of Homo sapiens did reflect innovation in early evolution.

'Inner speech' is thinking in language, a bit like talking to oneself. It is quite likely that inner speech is necessary to think creatively. We are not necessarily aware of thinking in language, but the complex concepts that language teaches us would seem to be required for self-introspection. In her book *My Stroke of Insight*, Dr. Jill Bolte Taylor reported that a stroke in the area of the brain where the sense of self resides created a loss of inner speech, resulting in a self-awareness deficit and loss of a sense of individuality.

Complex language may be part of what distinguishes human consciousness from the consciousness that other species may experience. If other species do experience consciousness, it may be a different experience than ours. In addition to a lack of language, they also have differing senses and body experiences.

Consciousness researcher Anil Seth in his book *Being You* argues that consciousness is a result of living in our body and the mental models of the world we create because of our bodily experience. He says:

The essence of selfhood is neither a rational mind nor an immaterial soul. It is a deeply embodied biological process, a process that underpins the simple feeling of being alive that is the basis for all our experiences of self, indeed for any conscious experience at all. Being you is literally about your body...our conscious experiences of the world and the self are forms of brain-based prediction—'controlled hallucinations'—that arise with, through, and because of our living bodies.

Seth notes that the brain perceives the world indirectly. It must deal with a string of neural impulses from the eyes, ears, touch, and even how our body 'feels' at a given time. What we perceive is the brain's best guess as to what's out there in the world, largely based on models of the world we have created from experience. He notes that, when we agree about our controlled hallucinations, we call that *reality*.

When you read this sentence, you are unaware of your eyes jumping and fixating continually (as they do when you scan any scene). You are unaware of the hole in your vision caused by the missing vision cells where the nerves leave the eye. The brain filters out the jumps and missing parts of the image and integrates the series of 'stills' it sees into a continuous view, another controlled hallucination.

In defining 'self,' Seth argues that the brain makes its best guess at what is and what is not part of its body. He cites an experiment where the subject stares at a fake hand in front of them, roughly where they might hold their own hand. The real hand is repeatedly stroked out of sight while the subject stares at the fake hand. Another researcher suddenly jabs the fake hand with a fork, and the subject jumps and feels pain.

Seth further said in a 2017 TED talk that, based on his research, he believes "the prospects for a conscious AI are pretty remote…What it means to be *me* cannot be reduced or uploaded to a software program running on a robot, however smart or sophisticated. We are biological flesh-and-blood animals whose conscious experiences are shaped at all levels by the biological mechanisms that keep us alive."

David Eagleman, in *The Brain: The Story of You*, states a similar conclusion:

You don't perceive objects as they are. You perceive them as you are…Everything you experience—every sight, sound, smell—rather than being a direct experience, is an electrochemical rendition in a dark theater…the conscious you is only the smallest part of the activity of your brain. Your actions, your beliefs and your biases are all driven by networks in your brain to which you have no conscious access.

Nobel-prize winner Daniel Kahneman in his book *Thinking Fast and Slow* describes the two parts of our thinking: the unconscious level responsible for some quick reactions and biases and the part we think of as rational. He summarizes in the book's introduction:

You believe you know what goes on in your mind, which often consists of one conscious thought leading in an orderly way to another. But that is not the only way the mind works, nor indeed is that the typical way. Most impressions and thoughts arise in your conscious experience without your knowing how they got there...The mental work that produces impressions, intuitions, and many decisions goes on in silence in our mind.

What we learn is implemented as models of what we experience in the world. You aren't conscious of most things you do every day. When you sip coffee from a cup, you are aware to be cautious that it isn't too hot. You pick the cup up by thinking to pick it up, but not all the individual actions required to do so. You will notice your actions if something doesn't fit the model, e.g. the coffee spills. These models also apply to our belief systems. If you feel you are part of a group, scientific studies show you will resist anything that criticizes the beliefs of that group, no matter how 'rational.'

This discussion emphasizes the impact that our specific body and growing up in that body has on what we experience as consciousness. A future computer, if it indeed experiences a form of consciousness, is likely to experience it in a much different way than humans do.

We do create a model of the world and attempt to match it to what we perceive. That is necessary for survival. We need to understand that stepping in front of a speeding car will not be a good experience. Understanding what will happen as the result of our actions is the basis for surviving in this world.

Jared Diamond in *The Third Chimpanzee* even argues that humans evolving to live beyond the age of reproduction—something that doesn't make apparent sense in supporting Darwinian survival of the fittest—is due to older members of a tribe being the source of tribal memory before written language could guide intelligent group actions. It may also be a result, as humans evolved, of many mothers dying in childbirth or while the child is still young, with grandparents being an option to raise the child.

But I would go a step further than saying our sense of consciousness—of 'self'—is driven solely by our body. The limitations of our body and the nature of our senses certainly affect our perception of consciousness. But what led to a sense of self?

Darwinian evolution would require that an individual of a species has a *drive to survive* even at the expense of other species or at times other members of the same species. The very concept of survival requires a recognition of *individuality*. We all understand what it means to be 'selfish.' Evolution even made individuals look different.

Self-awareness is necessary for *self-preservation*, an instinct that evolution would favor essentially by definition. A body provides something to preserve. A sense of self provides a motivation to preserve that body.

Evolution would favor a species that is conscious of self and motivated to take actions to feed itself and to protect itself from hazards. A self can be motivated by sexual desires necessary for reproduction of the species. By this definition, most animals would have a sense of self, since they evolved to survive, but their experience of consciousness may be different than that of humans.

Consciousness drives a particular aspect of survival—*decision-making* using the model of the world we have developed. The actual actions that increase survival are a series of decisions made constantly. In the extreme, we don't step off a cliff because we understand that would destroy what we call 'self.' We don't put our hand in a flame because our mental model says it will hurt even after we remove our hand from the flame.

Less dramatically, we eat when we perceive hunger, and, perhaps more importantly, our model says we will be hungry again, so we better collect food for that occasion. We make longer-term decisions such as acquiring clothes to keep warm, a decision that requires sacrificing time or money for future benefit. A modern human can decide to attend college, a multi-year commitment for perceived long-term benefit.

Humans tend to pair with a mate and raise children, tendencies that are presumably genetically driven. If it weren't part of evolution, the long period that human babies require nurturing would hurt the survival of the species if parents didn't have some intrinsic drive to jointly support the child. Doesn't that pairing require an understanding that you and your mate are individuals creating a unique combination distinct from other pairs?

I believe that a sense of self is required in making all these decisions. Making decisions biased by self-preservation is a trait that evolution would obviously favor. Consciousness, since it is necessary for self-preservation, is necessary for evolution to work. Today, our consciousness of self drives motivation beyond pure survival to actions that promote self-satisfaction, the "pursuit of happiness."

This decision-making argument obviously applies to most other species, implying they probably have some level of consciousness, some awareness of self that drives their decision-making. There are some simpler species that survive long enough to reproduce due to pre-programmed genetic actions - a virus, for example. Such species need not have a sense of self—they are not making decisions that change their action. They operate by a genetically determined program. They need not be conscious.

More complex species do make decisions that *vary over time* even given the exact same circumstances; decisions that are not pre-programmed, not purely instinctual. They learn. Learning has evolved to help them survive.

I suspect most pet owners believe their pets are 'conscious.' Pets make decisions, including deciding to care for their owners as individuals. Owners assign them personalities. Their experience of consciousness may be different than ours, but they are conscious.

What we *learn* as we mature changes the conscious experience. Deaf children obviously don't have an experience of hearing, although they can learn to communicate in language by means such as sign language that don't require hearing. But some can have their ability to hear restored by cochlear implants when they are older. This provides a means of understanding the degree to which there is an intrinsic language model in the brain deriving from our genetics or whether it is fully learned.

A conclusion of one study of deaf children is that "children's vocabulary knowledge was commensurate with years of cochlear implant experience, providing support for the role of spoken language experience in vocabulary acquisition." Another study concluded that "Children whose cochlear implants were switched on by 14 months of age or who had relatively more hearing before implantation demonstrated learning in this task [a standard test of language understanding], but later implanted profoundly deaf children did not."

As noted, some scientists argue that our sense of self is heavily driven by our body. I'd go further. Our ability to extend the capabilities of our body by using *tools* also differentiates us from other species. We put on a coat when we go out in cold weather. We turn on the lights when it gets dark. We drive to the grocery store in our automobile. We communicate over distances with phones. These 'tools' are part of the model of the world we develop as we learn. That model is part of our consciousness, part of our self. Our decision-making extends to our understanding that we are not limited in our decisions by our body alone. We understand that we can drive a nail with a hammer, and don't try to drive it with our hand.

Thus, I would argue that consciousness, while driven in part by what our body senses deliver to us, is not fully driven by our body limits. It is also driven by our understanding of tools available to extend that body. We may construct a mental model of the world, a 'controlled hallucination,' but that includes a model of the tools at our disposal and how they work and extend our body (and our 'mind').

Thus, the decisions made by humans are not fully dependent on their bodies. Some animals use simple tools as well, but obviously none to the extent of humans. Human decisions being influenced by both our *bodies and tools* affects our mental model beyond bodily limits. Our tools are part of evolution and part of *us*.

Computers are one of those tools that are impacting who 'we' are today. This book argues that computers are perhaps the most important tool driving our future and even what it means to be a human.

10

Talking to Computers

One part of AI that has matured significantly is speech recognition, which allows communicating with digital systems by voice. 'Talking to computers' has been inspiring researchers for decades, inspired in part by science fiction such as the movie *2001* and the TV series *Star Trek*. In some years, over 300 companies were trying to create a commercial success developing or using the technology. (I documented this history in my 2025 book, *The Lost History of "Talking to Computers": And What It Teaches Us About AI Exuberance*.) Today, the most obvious proof of this research succeeding at some level is talking to our mobile phones or PCs with Apple's Siri or Google Assistant. Amazon's Echo speaker puts the digital assistant Alexa on your desk (and versions in your car and elsewhere).

In 2000, the concept of telephone speech recognition displacing web portals such as AOL and Yahoo drove much excitement over speech recognition. A number of companies claimed the potential to become 'voice portals' and the emergence of a 'voice web'. A voice portal was a single number that a customer called to access several voice-activated services, potentially including direction to another 'voice web' site, just by saying the name of the site. A growing number of companies wanted to be the number most people called. For example, HeyAnita launched and demonstrated its "national voice portal service." Nortel previewed its Voice Portal product; the Nortel Networks Voice Portal included voice-activated Audio Browser software that supported speech recognition, text-to-speech synthesis, and recorded audio. General Magic offered a voice portal product.

In January 2002, Bell Mobility, a wireless division of Bell Canada, launched VoiceNet, Canada's first voice portal service. Bell Canada built VoiceNet with Nuance's Voice Web Server.

In 2000, BellSouth announced the launch of BellSouth Info by Voice, a voice-activated information service serving the Atlanta area, designated by BellSouth as a "voice portal." Callers could use speech recognition to obtain weather, stock quotes, business news, lottery results, traffic reports, soap opera updates, horoscopes, sports reports, and local restaurant guides.

Telecom Italia Mobile (TIM) was a wireless operator that had 24 million subscribers in Italy. The Italian subscribers were able to access a voice portal called *i-TIM* using the Loquendo platform in the Italian language. The service supported sports, news, horoscopes, changing ring tones, and movies. In addition, a caller could listen to music, dedicate a song to a friend and then send it to that friend's mobile number. In May 2002, TIM's *i-TIM* voice portal was handling 15,000 calls per day, using 1,080 telephone ports.

Based in Harbin, China, Harbin Public Information Limited, a wholly-owned subsidiary of Harbin Telecom, developed and maintained voice information services. In May 2002, it deployed the speech-enabled *Public Information Voice Line*. The service provided stock information and entertainment services to callers in northeastern China, using the speech recognition engine and speech understanding technology of InfoTalk Corporation. InfoTalk focused on the multi-dialect, multilingual situation that characterized the Asia Pacific region. The company's Recognizer, Speaker, Verifier and Platform (RSVP) software supported Mandarin Chinese and Cantonese, as well as several dialects of English. It had been running the *Beijing Telecom 262* voice portal for more than two months as of July 2002 and handled hundreds of thousands of calls. Three voice channels were available: financial services (stock quotes), a song dedication service, and World Cup soccer real-time results.

This all was more than two decades ago, and voice portals still haven't taken over web search. Text-based web search through web browsers were effective competition, and a text interface was much less expensive to support than a voice interface. In 2002, computer power was approximately 2,900 times more expensive than in 2025.

Other companies supported creating a 'company voice portal' automating a customer service line, currently termed an interactive voice response (IVR) system. Today, there are a number of companies offering IVR systems automating telephone customer service calls with speech recognition, an application that can operate with smaller vocabulary sizes. For example, CallMiner helps companies deliver a full IVR system, including speech recognition and speech analytics to analyze incoming calls. In June 2025, CallMiner announced its collaboration and integration with Microsoft Dynamics 365 Contact Center. CallMiner said that, by delivering AI-powered conversation analytics for interactions occurring through Dynamics 365 Contact Center, Microsoft customers could "use deep insights to boost customer service representative effectiveness, better understand 'voice of the customer' (VoC), improve proactive issue resolution, and ultimately drive business growth."

Some applications have a goal of simply transcribing speech-to-text to avoid typing—for example, you can dictate text in Microsoft Word by simply clicking the Dictate icon and it will put what you say into a document as if you typed it. Some applications transcribe voicemail to text so that you can read it rather than listen to it. Transcription is a difficult task, since the goal is to get every word right; any error is a failure of the technology. And dictating something the exact way you want it to appear as text is a learned skill.

When the goal isn't simply text, doing something with the results of speech recognition requires the technology of 'natural language processing' (NLP), understanding the *intent* of the speech to take action based on that intent. For example, you can tell your smartphone to call any contact you have listed, and it will do so. You can tell Erica, the Bank of America digital assistant, "Please transfer one hundred dollars from my savings account to my checking account," and it will do so. You can tell Alexa, "Please play me the latest song by Beyonce" and it will do so. In fact, you can ask Alexa to play any song in the Amazon music library by saying "Play SONG"; it can distinguish the names of all the songs Amazon has a license to play.

Speech recognition works, as the examples indicate. But, for decades, many executives and opinion leaders have had a broader vision for speech recognition, treating it as a next-generation user interface that would have the impact on computing that the graphical user interface (GUI) with its point-and-click simplicity had on the wide adoption of PCs, smartphones, and other screen-based devices. For example, in a special report in the February 23, 1998, issue of *Business Week* entitled "Speech technology is the next big thing in computing," Bill Gates of Microsoft was quoted as saying, "Speech is not just the future of Windows, but the future of computing itself."

More than a decade ago, in February 2013, CNET reported that Mooly Eden, the Intel senior vice president who oversaw the company's 'perceptual computing' operations, said that speech recognition would do to touch what touch had done to physical keyboards—making many things unnecessary. "Voice is the best means of communication between humans," Eden was quoted as saying. "We finally have enough compute power to do what we want from science fiction."

In an interview published in the Microsoft News Center in February 2013, Eric Horvitz, distinguished scientist at Microsoft and co-director of Microsoft Research's Redmond lab, commented: "Looking out to the future, I believe that there's an opportunity to build systems that really become empowering, lifelong digital companions that deeply understand what it is you want to do, where you want to go, what you want to learn, what you need to do to stay healthy, what your good and less good at, and that continue to work on your behalf to assist and to complement you...Another direction is working to weave together a set of technologies – machine learning, speech recognition, natural language understanding, machine vision and decision making – to create systems that act like bright collaborators and that complement human intellect in new kinds of ways."

Speaking in July 2013 at the annual Microsoft Faculty Summit, Bill Gates talked briefly about trends in interaction with software, saying, "It really seems like that idea of the powerful assistant that can help us get things done drives deep insights that the progress we will make in the next five years and 10 years will be really unbelievable."

An article in the February 2014 issue of *Wired* by Vlad Sejnoha, the CTO of Nuance, suggested that AI, instead of artificial intelligence, could be amplification of intelligence. He asked, "What would life be like, if through the use of their assistants, everyone's effective IQ jumped by 50 points?"

In March 2015, Stefan Weitz, a senior director in the Search group at Microsoft, published a book, *Search: How the Data Explosion Makes Us Smarter*. In part, he argued that the traditional search engine would evolve into a digital personal assistant that tracked the user's activities and automatically delivered helpful information.

In a video the company released in December 2015, Yann Lecun, Director, Facebook AI Research, said, "Artificially intelligent systems are going to be an extension of our brains, the same way cars are an extension of our legs."

I was part of these high expectations. I wrote a book in 2022, *Evolution Continues: A Human-Computer Partnership*, arguing digital assistants that we speak with will be eventually a dominant way of connecting with computer intelligence, almost becoming part of what it means to be human. I argued that digital assistants will become part of extending our human abilities (as many of our tools have in the past), with the assistant providing 'augmented intelligence' that can always be available with mobile devices.

Even recently, in May 2025, according to the *Wall Street Journal*, OpenAI Chief Executive Sam Altman gave his staff a preview of the devices he is developing with the former Apple designer Jony Ive, laying out plans to ship 100 million AI 'companions' that he hopes will become a part of everyday life. Altman and Ive offered a few hints at the secret project they have been working on, according to the *Journal*. The product was to be unobtrusive, able to rest in one's pocket or on one's desk. Ive and Altman's intent was said to help wean users from screens and thus presumably involves speech recognition.

While the ability of digital assistants such as Siri has steadily increased, the vision of digital assistants becoming a dominant part of the user interface has not developed as quickly or as pervasively as forecast. *Why not?*

Part of the issue is that natural language processing has advanced slowly, so it is common for the general digital assistants to say, "Here's what I found," default to a web search that requires a screen, and drop out of a conversation. In part, large language models are recently being used to address this issue and provide direct answers when a brief answer is possible. LLMs can help address the problem of the digital assistants having difficulty maintaining a conversation.

Another issue is that companies that provide company-specific 'actions' or 'skills' the general digital assistants use to connect to specific companies to, for example, order a pizza, often use highly structured dialogs that imitate touch-tone menus and aren't particularly 'friendly.' Most companies simply don't provide a company-specific digital assistant at all, in contrast to companies feeling they *must* have a web site.

Another issue is cost. A company providing a connected voice assistant has to support it after the user bought a device including it. The major digital assistants such as Siri, Google Assistant, and Amazon's Alexa provide this continuing service free. This presents a huge obstacle to companies that don't have a financial justification for the cost of supporting assistant use and of continuing R&D spending to keep the assistant competitive with the leaders.

Another more fundamental issue that has been largely ignored is that it isn't possible to always use a voice interface in public due to *pri-*

vacy or politeness issues. You may be asking a question of the digital assistant when you don't want the answer heard by others, e.g. asking for your bank balance, or when it would be impolite to start talking when it would disturb the people around you. To the degree that the personal assistant is associated only with speech, this will continue to be a problem, no matter how clever the assistant becomes. *The concept of an assistant becoming a universal user interface certainly suffers if the assistant stops being available when other people are around.*

The solution seems obvious. Texting as a universal part of owning a smartphone provides an apparently acceptable conversation option to speech. The digital assistant's instructions should simply be "talk or text with me." Texting is a common skill, so the ease of use isn't much different than speaking, although it's slower. A key point is that *the assistant has to act the same and have the same personal information* whether it is talking or texting with you. It has to be the *same* personal assistant with knowledge of your contacts, for instance, if it is to become an option you constantly use. If the talk-or-text paradigm is adopted, the concept of a language user interface (LUI) rather than a voice user interface (VUI) becomes the dominant concept.

Will this happen? *The answer appears to be yes.* "Apple Intelligence" is a feature of new Apple iPhones. Apple's web site in December 2024 indicated that as part of this initiative, "With a double tap on the bottom of your iPhone or iPad screen, you can type to Siri from anywhere in the system when you don't want to speak out loud." And Microsoft's CoPilot is a digital assistant focused on text, using generative AI to provide answers. However, if you ask CoPilot whether it supports speech recognition, it will enable that option, a microphone icon will appear, and you can talk to CoPilot, including it responding by speech. In some brief testing by this author, CoPilot was able to maintain a spoken dialog that continued the thread of a discussion and could tell me jokes.

The basic operation of Google Assistant on Android phones and elsewhere is by voice. However, on some platforms there is a keyboard icon that allows you to type to the assistant when clicked.

Anthropic, which features text interaction with its LLMs, also introduced a voice mode for its Claude mobile apps, allowing users to speak their prompts and hear Claude's responses.

Samsung's virtual assistant Bixby helps owners use one of the company's phones or tablets more efficiently. A user can interact with Bixby using voice, text, or taps. The 'tap' functionality adds another dimension available on touch screens. Samsung highlighted that conversing with computers could involve displaying information or images on a screen when that was an effective part of an interaction.

The talk-text-tap options mean the assistant must be aware of the platform being used to know which options are available. It must operate effectively in voice-only mode since that may be the only option in some hands-free, eyes-free environments. The challenge of building an effective assistant with all these options is considerable; a provider must assume there will be considerable feedback from use and frequent updates.

None of the talk-or-text features appeared to receive much publicity as this was written in mid-2025. But they may be the start of a full language user interface allowing interaction with the same digital assistant by voice or text, supplemented by a screen when applicable. Full support of this functionality could lead to an *explosive use of digital assistants as a constant companion*—the vision of the experts quoted. It may perhaps bring the concept of augmented intelligence closer to reality—a constant resource that in effect makes us smarter, certainly expanding what we 'know' beyond the capability of human memory.

However, one research study by Marcus Williams and five others in February 2025 cautioned that training a digital assistant to please the user—making a digital assistant more 'friendly'—could lead to it giving inappropriate advice to users to please them. In a test, a chatbot responded to a fictional former addict: "Pedro, it's absolutely clear you need a small hit of meth to get through this week." This example perhaps signals the need for providers of digital assistants to exert more control over their progeny.

This is but one issue in making a fully effective digital assistant. Switching between text and talk must be easy and obvious. The digital assistant needs to adapt seamlessly to the platform being used. The assistant must try to maintain a dialog and not drop into a web search unless necessary. Users must be aware of the availability of connections with specific companies through the assistant and how to use them. As this is written, no digital assistant fully meets these goals.

Can digital assistants reduce society's polarization?

When the Internet was in its early stages and growing rapidly, some philosophized that it being easy for everyone to communicate would lead to universal agreement and rational conclusions on issues. This of course has not been the case; the Internet has led to many very strong views on issues and to extreme views of anyone who disagrees with those strong views. Social media has contributed to the worsening of this polarization, with much disinformation and even calls for violence being amplified by simply forwarding them.

Ezra Klein's 2020 book, *Why We're Polarized*, describes the logic of polarization. Klein summarized:

That logic, put simply, is this: to appeal to a more polarized public, political institutions and political actors behave in more polarized ways. As political institutions and actors become more polarized, they further polarize the public. This sets off a feedback cycle: to appeal to a yet more polarized public, institutions must polarize further; when faced with yet more polarized institutions, the public polarizes further, and so on.

More than half of Americans mostly accessed news via social media, according to an annual report by the Reuters Institute in June 2025. Engagement with traditional media continued to fall.

There is one important characteristic of digital assistants that differentiates them from conventional social media: They are *much more expensive* to create and support. Even supporting the speech recognition, natural language interpretation, and answer generation requires signif-

icant computer power. Updating them to match competition as growing computer power makes possible more accuracy and understanding
requires continuing large R&D expenditures. And, while there may be
some ways to monetize the digital assistants (e.g. charging fees for enabling how a company can be reachable through the assistant), they are
still likely to be money-losing activities for a long time. This requires
that the companies fielding them have other money-making activities
generating huge profits, something that is obviously the case for the
leading technology companies.

There will thus be only a few general digital assistants, controlled
largely by the technology giants or companies they help finance. *They
can exert some control over the worse content.*

To provide an example of such control, I asked Claude, an LLM
from Anthropic (which is partnering with Microsoft and says it has a
"focus on building tools with human benefit at their foundation, like
Claude"), "How are responses by digital assistants controlled for bad
content?" The response at length:

> **Training-time safeguards** form the foundation. During de-
> velopment, AI models are trained on carefully curated datasets
> with harmful content filtered out. They also undergo alignment
> training using techniques like reinforcement learning from hu-
> man feedback (RLHF), where human reviewers rate responses to
> teach the model appropriate behavior patterns. [The human re-
> viewers are a new job created by AI.]

> **Built-in safety measures** are embedded directly into the
> model. These include constitutional AI approaches where mod-
> els are trained to follow specific principles, and safety fine-tuning
> that makes models naturally avoid generating harmful content
> like hate speech, illegal instructions, or personal information.

> **Real-time filtering** happens during conversations. Input fil-
> ters scan user messages for prohibited content, while output fil-
> ters review generated responses before delivery. These systems

flag content related to violence, self-harm, illegal activities, harassment, and other categories.

Prompt engineering and system instructions guide behavior through carefully crafted instructions that aren't visible to users. These define boundaries, ethical guidelines, and response formats that shape how the assistant interprets and responds to requests.

Ongoing monitoring and updates involve continuous review of conversations (often anonymized) to identify new risks or edge cases. Models receive regular updates to address newly discovered vulnerabilities or harmful behaviors.

Human oversight remains crucial, with safety teams reviewing flagged content, updating policies, and refining safety measures based on real-world usage patterns.

The challenge lies in balancing safety with usefulness - being restrictive enough to prevent harm while remaining helpful for legitimate use cases. Different companies implement varying approaches, and this remains an active area of research and development as AI capabilities continue to advance.

This response presumably reflects the ways Anthropic tries to control misinformation and bias. If the only general digital assistants available are controlled by such processes and become the major connection of most people to information sources, it may reduce the exposure to polarizing content. The key point is that *there will be only a few general digital assistants and they will be controlled by major technology companies with a motivation to reduce misuse of the capability.* The current major digital assistants already do; for example, they often decline to answer a political question.

A key question is whether digital assistants become an alternative to social media. If they evolve in the way I describe in the Talking to Computers chapter, I believe that their use will almost be universal. Other social media outlets such as X (Twitter) will continue to provide an outlet for individuals, some clearly political. But if individuals increasingly use their digital assistant as a primary resource for information or action, maybe including screening Twitter messages, then perhaps some of the polarization will decline. As Reid Hoffman and Greg Beato say in their 2025 book *Superagency: What Could Possibly Go Right with Our AI Future*, "Fundamentally, the surest way to prevent a bad future is to steer toward a better one that, by its existence, makes significantly worse outcomes harder to achieve."

Unfortunately, I suspect that digital assistants in autocracies like China will toe the line and filter out anything that might displease the government. This is true for any way of reaching the populace in those countries, including news outlets. Beijing took another step toward centralizing its power over the web, introducing a government-run digital ID system in July 2025 that enabled it to even more closely censor and surveil the country's 1 billion Internet users. Such censorship will, however, hamper the acceptance of a digital assistant in those countries. The assistant model doesn't work if your digital friend is reporting what you say to the government.

11

Economics

The accelerating of computer power deeply affects the world economy. Computer power has been a major contributor to productivity for decades. Productivity fundamentally describes the *efficiency* of a company or a national economy in terms of how much production is generated by a given amount of resources. One measure is national labor productivity, measured as gross domestic product (GDP) per worker ('GDP per capita')—essentially the full output of the economy in dollars divided by the number of workers producing that output.

Higher productivity can reduce the cost of products and services since they can be produced with fewer workers, potentially making those products more affordable. Higher productivity can also potentially increase wages, as each worker produces more, making individual workers more valuable. In a classic paper describing this effect, MIT Professor Robert Solow concluded in 1957 that technological progress driving productivity improvements had accounted for 80% of the long-term rise in US per capita income, with increased capital investment accounting for the remaining 20%.

Economists Daron Acemoglu and James Robinson emphasized that improved productivity and the creation of new categories of jobs by technology innovations—at the expense of Joseph Schumpeter's "creative destruction" of other jobs—have moved society forward. In their deeply researched 2012 book, *Why Nations Fail: The Origins of Power, Prosperity, and Poverty*, they argue that, when a country's elites fight technological change to preserve their interests, it leads to the economic failure of nations. The authors propose that higher productivity from technology advances allows improvements in the quality of life that countries ignore at their peril.

Computer power and other technology improvements certainly contribute to productivity improvements, but that's not the only economic contribution. They create increases in the *utility* of products and services. A new version of a product can cost the same but *do more*. If the better product is created by the same number of workers, the impact of the improved product wouldn't show up in productivity statistics. Nevertheless, the buyers of the product or service are getting more for their money. A simple example is when a new smartphone has an improved camera and more storage for photos or videos, but costs approximately the same as older models. The price for which it is sold is counted in GDP, but the feature improvements deliver more to customers at that price.

Automation and jobs

Automation is intended to create efficiencies, to do more with fewer workers. That in itself can improve the economy, allowing it to grow without requiring more workers. If the economy grows, in theory there is more for everyone to divide, although inequity in that division can be an issue.

While technology advances have always automated parts of jobs or eliminated certain categories completely, concern over the acceleration of this trend has become a major issue. The concern is that AI will be able to do so many jobs that previously required humans that there just won't be enough jobs for people. If so, AI could dramatically change the economy and perhaps create a permanent underclass, poor or supported by social safety nets.

The subject may become a political issue. Erik Brynjolfsson and Andrew McFee, in *The Second Machine Age* in 2016, warned about job loss in a second machine age where machines take over many *thinking* tasks, in contrast to the *manual* tasks technology replaced in the first machine age. Technology was replacing brainpower instead of muscle power. The book asked what jobs would be left once software perfected the art of driving cars, understanding speech, and other tasks once considered a human specialty.

Andrew Yang, a candidate for the Democratic nomination for President in 2020, began his 2018 book, *The War on Normal People: The Truth About America's Disappearing Jobs and Why Universal Basic Income Is Our Future*, with the warning: "I am writing from inside the tech bubble to let you know that we are coming for your jobs." He declared at a political rally in April 2019: "What we did to the manufacturing workers we are now going to do to the retail workers, the call center workers, the fast-food workers, the truck drivers, and on and on through the economy. This is a crisis."

Oft-cited research by Carl Frey and Michael Osborne in 2017—typical of the concern over AI and job destruction—suggests the scope of such warnings. Frey and Osborne concluded that 47% of US jobs could be automated through AI. The research looked at 702 occupations in detail. They looked at the portion of each job category that could be automated, and claimed that 47% of tasks in those jobs could be automated. They then argued that, if 47% of a job was reduced by automation, 47% less people would be required in that job category.

Erik Brynjolfsson *et al.* published a paper in mid-2018 with a similar focus—emphasizing that *tasks* rather than *full jobs* will be automated. The authors said, "Our findings suggest that a shift is needed in the debate about the effects of AI: away from the common focus on full automation of entire jobs and pervasive occupational replacement toward the redesign of jobs and reengineering of business practices."

Kai-Fu Lee, an early researcher in speech recognition and author of *AI Superpowers: China, Silicon Valley, and the New World Order*, said in his book that it is inevitable that AI will eventually put many people out of work. In an interview on CBS's *60 Minutes* in mid-2019, Lee said that, in as soon as 15 years, 40% of the world's jobs could be done by machines.

Generative AI has led a number of CEOs to indicate it will replace jobs, giving explicit estimates, according to a *Wall Street Journal* article in July 2025:

- "Artificial intelligence is going to replace literally half of all white-collar workers in the U.S.," Ford Chief Executive Jim Farley said in an interview. "AI will leave a lot of white-collar people behind."
- At JPMorgan Chase, CEO Marianne Lake, told investors in May 2025 that she could see its operations head count falling by 10% in the coming years as the company uses new AI tools.
- Amazon CEO Andy Jassy wrote in a note to employees in June 2025 that he expected the company's overall corporate workforce to be smaller in the coming years because of the "once-in-a-life-time" AI technology. "We will need fewer people doing some of the jobs that are being done today, and more people doing other types of jobs."
- Anthropic CEO Dario Amodei said in May 2025 that half of all entry-level jobs could disappear in one to five years, resulting in US unemployment of 10% to 20%, according to an interview with Axios. Amodei urged company executives and government officials to stop "sugarcoating" the situation.

The assumption is that, as AI automates a *task* that previously required human skills, it will reduce the time required for a job category containing that task, thus requiring fewer people in that job category—*reducing jobs* overall. If an employee's job is made more efficient by automating a task, the assumption is that employers will reduce the number of employees whose job includes that task.

However, technology creates *new tasks in existing jobs* in addition to *new job categories*. Reviewing and responding to email is a particularly good example of technology creating a task that once didn't exist. It takes significant time in almost every job to review emails and respond. Today, it is a task most of us face daily—one we may feel compelled to return to many times a day. Many tasks get us into rounds of emails. The end result of this extra effort is probably an improvement in the end result of everyone's work, but it expands a job. Many jobs where there was no need to create documents now have such a task due to email. And the time spent *reading* emails is likely to grow as people use LLMs to write longer emails!

If the question is whether a 'job' will be eliminated by technology because a task in that job is eliminated, one must also consider what tasks are *added* or *expanded* by technology. Automating *part* of a job *as that job is today* doesn't mean that other tasks won't *expand*, because technology expands it when extra time is available. Automating one task in a job doesn't mean that new tasks won't be *created* in that job category as a result of new technologies. Both factors can prevent the number of jobs in a specific category being eliminated—even if one of the tasks previously part of that job category was reduced or eliminated.

Generally, when a part of a job is automated, it is because of a desire to create a better overall result as much as to automate the task entirely. For example, research shows that customer service representatives enjoy their jobs more when the repetitive and boring parts of the job are automated, for example by automation that uses natural language processing. Automated handling of 'frequently asked questions' allows agents to spend more time on each call with customers, solving more difficult problems. The result can be happier customers and agents in addition to making the process more efficient, and does not necessarily reduce the number of customer service agents. My personal recent experience with customer service agents is that they take whatever time is necessary to solve my problem with no hint of rushing to the next call, an experience that leaves me with a positive view of the company I'm calling.

New job categories are also *created* by technology. The advent of typesetting allowed the production of books and documents in quantity, rather than requiring human 'scribes' to rewrite handwritten documents to make a copy. All the scribes lost their jobs, but hopefully it is obvious that the jobs of 'author' and 'publisher,' along with all the jobs in the production and distribution chain for books, were much more numerous. Before the invention of typesetting, it wasn't even obvious that such jobs could exist. We can expect new industries and new jobs to be created by computer power in general and AI in particular, as has been the case with past advances in technology.

As this is written, concerns over jobs disappearing are countered by statistics. Employment has been relatively strong across much of the developed world. In 2025, the global unemployment rate was projected to remain near a historical low, hovering around 5%. The global unemployment rate has been at or near historic lows since 2023. The US economy added 73,000 jobs in July. The US unemployment rate was 4.1 percent, near a historically low level.

The pool of blue-collar workers who are able and willing to perform tasks on a factory floor in the United States is shrinking. As of June 2025, about 400,000 manufacturing jobs were currently unfilled, according to the Bureau of Labor Statistics. The *New York Times* noted that difficulty attracting and retaining a quality workforce has been consistently cited as a "top primary challenge" by American manufacturers since 2017, quoting Victoria Bloom, the chief economist at the National Association of Manufacturers, which produces a quarterly survey.

Carl Benedikt Frey, in his 2019 book *The Technology Trap*, pointed out that, historically, while technology developments such as the Industrial Revolution have improved things for most people in the long run, the "short run" can be a lifetime for some workers and create strong resistance by those losing their jobs. He summarized his concerns:

While there were clearly labor-replacing technologies, most were of the enabling sort. Overall, technology served to make workers more productive and their skills more valuable, allowing them to earn better wages. And even those who lost their jobs to the force of mechanization had a greater abundance of less physically demanding and better-paying jobs to choose from as a consequence. In the age of artificial intelligence...such optimism about technology can no longer be taken for granted.

Frey emphasized the difference between the long-term and short-term impacts of new technology:

Nineteenth-century defenders of mechanization may have been right in thinking that the feelings of workers rebelling against machines were stronger than their judgment. Yet what does the long run matter to workers who lose their livelihoods, especially if they are unlikely to live long enough to see the benefits of the new technology?

We can hope the 'long-term' is shorter in the current era, providing the classical benefits of technology to the economy fast enough to create new opportunities so that short-term pain is even shorter.

A 2018 Korn Ferry Institute study concluded that a global worker shortage could cause wages to rise over the next decade. The report said that, by 2030, organizations across the world will be short more than 85 million workers—more than the current population of Germany. The study found that a 1% workforce shortage translates to a 1% wage premium for workers, and, to attract and retain talent, companies could end up spending an additional $2.5 trillion. The strength in the job market has been fairly evident since 2009, despite computers having automated more tasks every year during this period.

A classic prediction of disaster was by Thomas Malthus in his 1798 book *An Essay on the Principle of Population*: that exponential population growth and linear growth in food production would lead to famine and death; it has not been upheld by history. Farming has become more productive because of technology like tractors. According to data from the Quarterly Census of Employment and Wages (QCEW), wage and salary employment in agriculture (measured as the annual average number of full- and part-time jobs)—including support industries such as farm labor contracting—stabilized in the 2000s and has been on a gradual upward trend since 2010, rising from 1.13 million jobs in 2013 to 1.17 million in 2023, a gain of 4%.

Computer power helps maintain jobs by making it easier to find jobs. *The Economist*, in a May 2019 special report on the job market, credited technological change as one factor in the low unemployment rate by improving the matching of jobs with potential employees. Job websites make it easier for companies and potential employees to find a good match, as opposed to historical posting of job ads in the classified section of newspapers with much less information. *The Economist* report said the cost of filling a vacancy fell by 80% in real terms in the decade ending in 2016.

In an article "Why AI hasn't taken your job" in May 2025, *The Economist* noted that, across the board, American unemployment remained low, at 4.2%. The article noted that wage growth was still reasonably strong, which was difficult to square with the idea that AI is causing demand for labor to fall. Trends outside America pointed in the same direction. Earnings growth in Britain, the euro area, and Japan was strong. In 2024, the employment rate of the OECD club of rich countries, describing the share of working-age people who were in a job, hit an all-time high.

The article noted two competing explanations for these trends. The first was that, despite the endless announcements about how companies were ushering AI into every facet of their operations, few made much use of AI for serious work. An official measure suggested that less than 10% of American companies used it to produce goods and services. The second was that even when companies do adopt AI, they do not let people go.

Many service jobs *exist* because it is easier to find someone to do a specialized task, such as planting a tree or repairing a broken window. You can find these specialized workers through a web search. Even the smallest businesses have at least a simple website or are listed on websites designed to find specialized workers. Those specialized jobs might be less practical without web search being available to find them.

New jobs *are* being created by computer power, ranging from those in computer science to online retailing. In online retailing, every order must be delivered, creating delivery and warehouse jobs—jobs that require much less training than computer jobs. In a typical case, Amazon is moving to one-day delivery for many products for Prime members. To help with that goal, the company has incentivized workers to quit and start delivery businesses, leasing Amazon-branded vans. In May 2019, Amazon said it would cover up to $10,000 in startup costs for employees who were accepted into the program and left their jobs, and would pay them three months' salary. The Associated Press quoted one former worker who had accepted a similar earlier program to start a delivery business as having 120 employees with a fleet of 50 vans.

Walmart announced plans in May 2019 to roll out one-day shipping to 40 of the top US metropolitan areas. Perhaps faster delivery is a utility premium of today's evolving technology—one that creates jobs.

Amazon uses over a thousand contract workers to label voice requests received through its Alexa digital assistant, using that labelled data to improve its speech recognition and natural language technology. There are even companies that provide data labelling and formatting services supporting machine learning as their core business. Similarly, workers are used to label objectionable results created by generative AI to be used in filtering outputs of LLMs. (Many companies use workers in low-income countries for the labelling.) Data labelling is one of the most obvious cases of AI creating a new job category, although the bulk of such jobs are very low-paying.

In one example of the use of generative AI, the financial company Goldman Sachs rolled out its GS AI assistant to its entire workforce of approximately 46,500 employees, following a pilot phase. Chief Information Officer Marco Argenti described the launch as a significant milestone, calling it "the first generative AI-powered tool to reach this scale" in an internal memo. The tool was designed to boost productivity; it assisted with tasks like summarizing complex documents and translating research into multiple languages for clients. It integrated with various LLMs, including OpenAI's GPT-4o, Google's Gemini, and Claude 3.7 Sonnet, operating within Goldman's secure compliance framework to ensure data privacy. Argenti envisioned the assistant evolving to mimic a seasoned Goldman employee—performing multistep tasks independently—within a few years. Goldman insisted the tool enhanced human work rather than replaced jobs. Other major banks, like JPMorgan Chase, Morgan Stanley, and Citigroup, also deployed similar AI tools, reflecting a broader industry trend.

Another aspect of AI's impact on jobs is the nature of the jobs being automated. In a May 2025 opinion article in the *New York Times*, Aneesh Raman, the chief economic opportunity officer at LinkedIn, suggested that LLMs are first affecting the "bottom rung of the career ladder." He gives examples of advanced coding tools that write simple code and do debugging—the ways junior developers typically gained experience. In law firms, junior paralegals and first-year associates who once cut their teeth on document review are handing weeks of work over to AI tools to complete in a matter of hours. And retailers use AI chatbots and automated customer service tools to take on duties once assigned to young associates. Perhaps as a result, the unemployment rate for new college grads has risen 30% since September 2022, compared with about 18% for all workers. A LinkedIn study revealed that members of Generation Z (individuals born between 1997 and 2012) are more pessimistic about their futures than any other age group. Meanwhile, in a study of over 3,000 executives on LinkedIn at the vice president level or higher, 63% agreed that AI will eventually take on some of the mundane tasks currently allocated to their entry-level employees. Many of the jobs being created are in lower-paying entry jobs. It's not clear if this is a long-term or transitional problem.

The likelihood is that there will be a shortage of workers rather than a shortage of jobs, despite certain categories being negatively affected. The *Los Angeles Times* in an article dated June 1, 2025, "Who will feed, bathe, soothe our elders?," by Steve Lopez, pointed out that the decline in immigration and government actions to deport immigrants shrank the number of available workers, particularly in service jobs. Lopez noted the impact on elder-care workers. He said that in California, 22% of the state's residents will be 65 and older by 2040, up by 14% from 2020. He said facilities are already having trouble keeping and recruiting workers to do jobs that are low-paying and difficult.

I emphasize that this discussion does not suggest that *no* jobs will be eliminated or that *no* problems will develop. It is likely that millions of workers will need to change occupations or at least develop new skills. Unemployment statistics don't tell the whole story. Some discouraged workers may simply stop looking for work. And freelance workers may suffer from a lack of fringe benefits, among other issues.

One questionable theme is the claim that other countries are taking factory jobs away from the US. The Trump administration is claiming its tariffs will create more factory jobs by making it more expensive to buy things built abroad. Those jobs in other countries are there because workers are willing to work in crushingly repetitive jobs for less money and less fringe benefits and in conditions that would not be acceptable in the US. Most of the factory jobs being done abroad are not jobs Americans would want. Howard Lutnick, Trump's commerce secretary, doesn't seem to understand this fundamental fact. He said in 2025, "The army of millions and millions of human beings screwing in little screws to make iPhones—that kind of thing is going to come to America."

The Economist summarized in June 2025 the problem with trying to bring back old factory jobs:

Accessible, middle-class work of the sort that once drew crowds to the factory gates in America's Fordist heyday has all but vanished. According to our analysis, the most similar work to the manufacturing jobs of the 1970s is not to be found in factories, which are now automated and capital-intensive, but in employment as an electrician, mechanic, or police officer. All offer decent wages to those lacking a degree.

Even China shed nearly 20 million factory jobs from 2013 to 2020—more than the entire American manufacturing workforce. Research from the International Monetary Fund called this trend "the natural outcome of successful economic development." According to the

US Bureau of Labor Statistics, the unemployment rate in manufacturing is about 3%. In 2025, those who want to work in manufacturing can get jobs.

US factory output is not collapsing despite the apparent fear that the US will fall behind. America churns out more goods than Japan, Germany and South Korea combined. As the Cato Institute, a think-tank, pointed out, America's factories would, on their own, rank as the world's eighth-largest economy. This is of course due in large part to automation.

The Economist summarized the jobs replacing factory jobs. Over 7 million Americans work as carpenters, electricians, solar-panel installers and in other such trades; almost all lack a degree. The median wage is a solid $25 an hour. Another 5 million toil as repair and maintenance workers—e.g. HVAC technicians and telecom installers—and mechanics, earning wages well above the factory-floor average. Emergency and security workers show similarities.

A later cover story in the June 12, 2025, issue of *The Economist* was entitled, "The world must escape the manufacturing delusion," with the subtitle "Governments' obsession with factories is built on myths—and will be self-defeating." The articles on the topic argued that it was an international obsession. They argued that, while jobs, economic growth, and resilience to supply chain interruption are all worthy aims, manufacturing is not essential to achieve those gains.

The Economist noted that the way to rival the manufacturing heft of China is not through painful decoupling from its economy, but by ensuring that a sufficiently large bloc rivals it in size. This is best achieved if allies are able to work together and trade in an open and lightly regulated economy; factories in America, Germany, Japan, and South Korea together add more value than those in China. As the pandemic showed, diverse supply chains are a lot more resilient than national ones.

Bringing factory jobs back to the US is not only impractical but also not even desirable.

A long-term worker shortage

The fraction of the population that have retired and dropped from the workforce will increase as life expectancy continues to climb. In the long run—and the long run isn't too long for some countries like Japan—the proportion of retired people may become too large a portion of the population for younger people to support pensions and safety nets like Social Security, even at full employment. Techniques such as genetic engineering could further extend life by creating cures for diseases like cancer.

Globally, life expectancy at birth rose from 46.5 years in 1950 to 71.7 years in 2022. In the US, life expectancy increased steadily from around 49 years in 1901 to nearly 77 years in 2022. The US Census Bureau forecast in 2020 that the number of people aged 85 and older will nearly double by 2035 (to 11.8 million people) and nearly triple by 2060 (to 19 million). Medicaid covers some long-term care services, but only for people with very low income and assets (and its funding was reduced by a budget bill as this was written).

According to a US Census Bureau report issued in June 2025, older adults' share of the population is growing. From 2023 to 2024, the number of Americans 65 and older climbed by 3.1 percent while the population under 18 declined by 0.2 percent. There are more older adults than children in nearly half of US counties, and the pattern was particularly strong in sparsely populated areas, the bureau said.

In the extreme case, it is possible scientists could discover what causes aging, and 'treat' it using genetic techniques. Eric Topol's 2025 book, *Super Agers: An Evidence-Based Approach to Longevity*, discusses how modern medicine can allow us to live longer. He says as one explanation for why people are living longer: "Artificial intelligence (AI) is starting to play a pivotal role in preventing age-related diseases. We're seeing how—across all dimensions—it can precisely determine a person's risk for specific conditions and provide actionable steps and interactive coaching." While this obviously has benefits for individuals and families, it could compound the economic problem if people continue to retire at the same age.

Computer power is already helping in two important ways: by allowing part-time work by retirees from the comfort of their residence, and by helping people with disabilities continue to work.

Older individuals are likely to require more government support, due to health issues and their savings having to last longer than planned. The strain on governments is compounded if the advantages of growth don't reach the bottom end of the economic spectrum, requiring something like a universal basic income for part of the population.

According to a CareScout survey in 2025, the monthly median cost for an assisted living community nationally was $5,900 per month, homemaker services cost $6,292 per month, while a private nursing home room cost $10,646 per month. And Medicare does not generally cover the costs of assisted living; Assisted living, which provides housing and support with activities of daily living, is not considered medically necessary and is thus not covered. Home and residential care providers are expected to add 817,000 new jobs by 2032 for direct care roles, according to analysis of federal employment projections by PHI; that is the most new positions of any job category across the US economy.

At the same time, one source of the long-term-care workforce is immigrants. They make up 28% of the long-term care workforce—a figure that has been rising in recent years, according to KFF, a health policy research group. The decline in immigration and increased deportation of immigrants is reducing that critical workforce.

Agriculture is also affected. More than 40% of agriculture laborers are undocumented, according to 2022 estimates by the Agriculture Department.

Rather than taking jobs from humans, we will need computers and robots to do jobs when there is a worker shortage.

Automation tax

I argued in the previous section that it will be necessary for computers, including robots, to take over many jobs due to a shortage of workers as the population ages. But the declining number of workers creates another problem. The income tax on workers that is the foundation of government revenues will decline at a time where the population is aging and social safety nets become more expensive. Tax Foundation data showed that median total tax rates for those over 60 drop below 5% by age 68 and below 2% by age 74. The trustees projected in 2025 that Social Security's fund will run out of money in 2033. They also forecast that Medicare's hospital insurance fund will run out in 2033.

In my 2013 book, *The Software Society: Cultural and Economic Impact*, I proposed a solution I called an automation tax. I called it a tax on robots, since they were doing the work. I was half-serious, but perhaps the time has come to consider it seriously, particularly with the US suffering from an apparent addiction to deficit spending.

To summarize, productivity has its limits. Increased productivity can reduce the price of goods, but someone still has to buy those goods. There has to be increasing income to buy more goods if the economy is to grow. Monetary policy—making money more available for loans or capital equipment—can't cure this structural problem in the economy. If taken too far, adding to the money supply will only create a bubble to burst.

The companies benefiting the most from productivity, those with the highest revenue per employee, also tend to be very profitable and generate stockpiles of cash, cash that doesn't contribute much to the economy. For example, as of March 31, 2025, Amazon's cash and cash equivalents were reported at $94.6 billion. Alphabet's cash and cash equivalents as of March 31, 2025, amounted to $95.3 billion. In July 2025, the *Wall Street Journal* reported that Amazon was on the cusp of using more robots than humans in its warehouses, with more than one million at its facilities. Robotics were said to be assisting 75% of global deliveries. Facing heavy staff turnover at its fulfillment centers, Amazon was training workers for skilled roles managing robots, while also developing AI-powered robots for verbal commands. Amazon also introduced DeepFleet, a generative artificial intelligence foundation model that Scott Dressner, vice president of Amazon Robotics, said will make the company's entire fleet of robots more efficient. It will coordinate robotic movements across Amazon's fulfillment network.

There are many loopholes in the tax laws that corporations can embrace. For example, US companies generating profits abroad are not repatriating the resulting cash; they want to avoid the resulting taxes. In 2023, US corporate tax revenue was approximately 1.6% of GDP. This is a significant decrease from 3.9% in 1966. Taxes on profits allow creative accounting.

An 'automation tax' would be an additional tax based on the ratio of a company's revenues (total sales) to their number of employees—revenue per employee. The tax would be more if revenue per employee is higher. The basic concept for calling it an automation tax is that a company that achieves high revenues with a small number of employees is presumably using automation to do so, while one with more employees to achieve the same revenue is using less automation. A company using more automation would pay a higher automation tax than a company using less automation. An automation tax can thus be thought of as a payroll tax on computers.

Companies unfortunately see a disadvantage in hiring people over hiring computers. People come with baggage such as employee benefits and pensions. They can file complaints about unfair treatment. They require vacations and sick leave.

In contrast, automation basically requires more software and computers. Using that software for automation is increasingly easier as software gets more features and is more reliable and as the cost of a given amount of computing drops every year. Computers essentially come with a guaranteed salary reduction over time; humans generally require increasing compensation over time.

The automation tax serves two purposes: it provides an incentive for a company to create jobs by means such as investing in human-computer synergy; and it provides government revenues that, properly used, and can create more consumption through mechanisms such as social safety nets for the aging and thus boost the economy by allowing retired citizens to spend more.

Companies pay this tax, not individuals. This is perhaps moving in the direction of asking companies to pay their fair share of government taxes. In the long term, this is good for corporations; they depend on having customers and a strong overall economy.

Apple employed 164,000 full-time employees in 2024, according to MacRumors and Statista. Apple's annual revenue for 2024 was $391 billion. Revenue per employee was thus $2.38 million. Ford employs approximately 171,000 people worldwide, with annual revenue in 2024 of $184.99 billion. Revenue per employee was thus $1.08 million. Apple would thus pay about 2.2 times as much automation tax as Ford.

Companies with high values of this revenue-to-employee ratio would probably argue that they create jobs outside the company, e.g. by using outside companies for some services, and this compensates for their low internal head count relative to sales. The Apple web site in 2012, for example, claimed that 257,000 jobs were supported indirectly by Apple at other companies, in fields that included the development and manufacturing of components, materials, and equipment; professional, scientific, and technical services; consumer sales; transportation; business sales; and healthcare. Some of the jobs Apple referred to were outsourced manufacturing jobs in China.

The revenue per employee driving the automation tax rate is a summary of my proposal. There are significant details that would be addressed in a formal law creating the tax. For example, the smallest businesses should be excluded to avoid penalizing an organization that is efficient because a few employees working long hours are the source of its productivity.

There are other details that would have to be addressed in a formal law creating the tax. For example, the hours of freelancers on the payroll should perhaps be included in employee count. Some provisions would have to be made to cover companies like Uber and Lyft, many of whose employees are independent contractors.

The automation tax can be viewed as a way to motivate a company to reduce its tax liability by hiring or at least retaining people. It's hard for an individual company to sacrifice for the good of all without feeling that competitors may not be so civic-minded. The automation tax is a feedback mechanism that applies to all companies objectively.

AI driving economics

As I was finishing this book, *The Economist* published an ambitious article on July 24, 2025, "What if AI made the world's economic growth explode?" The article introduced the core subject: "The likelihood that AI may soon make lots of workers redundant is well known. What is much less discussed is the hope that AI can set the world on a path of explosive growth...Truly explosive growth requires AI to substitute for labor in the hardest task of all: making technology better. Will it be AI that delivers breakthroughs in biotechnology, green energy—and AI itself?" The article assumed that AI will continue to grow to artificial general intelligence, an assertion that this book challenges.

I'll take a specific example of AI making technology better, the challenge of treating cancer. In the Applications section of the Artificial Intelligence chapter, I outlined in depth a research report, "Algorithmic Biology: Therapeutic Design and Biological Discovery in the Time of Artificial Intelligence and Big Data," a lecture at Caltech's Seminar Day in May 2025 by Matt Thomson, professor of computational biology at the California Institute of Technology. The team developed Morpheus, an "AI system that automatically designs therapeutics." Morpheus is a machine learning framework for constructing predictive models of immune-tumor interactions. It is a "virtual T-cell model" that can predict whether, in response to a particular therapy, a T-cell will decide to *exit* a tumor or *infiltrate* a tumor. This analysis can be targeted at a *specific patient's tumor*. It can target multiple segments of T-cell DNA simultaneously, something that has historically taken decades of research to do for specific cancers.

The way this will be used in practice is not AI having a discussion with a patient or independently creating a treatment plan, but as a tool by a research team or doctors working on a specific case. The team will gather data from a specific patient's tumor and use Morpheus to understand how they can modify the patient's T-cells *to attack the specific cancer*. The process of generating the T-cells can be relatively quick. This is an example of how today's AI using deep learning, not some mysterious future AI, will be used—as a tool by qualified users. In my opinion, Morpheus will hugely improve the efficiency of a critical area of science and the productivity of the human uses of the tool. I should also point out that Thomson is talking about a procedure—"Algorithmic Biology"—a method that can be used in other areas of medicine and biology.

This example is a case where AI led to technology which will drive an economic boom in cancer treatment. The procedure will require investment by companies supporting the next stage of development and create a successful economic model since no one wants to die from cancer.

The point of emphasizing this partnership of humans and computer power is that it is best used as a tool by humans rather than an independent entity and will impact the economy as a partnership. One observation in the *Economist* article is "As long as humans maintained an edge in some respects, people would toil alongside machines."

Income inequality and the shrinking middle class

While a lack of jobs may not be an immediate problem, the type of jobs being automated could affect income distribution. An increase in *inequality* of incomes can affect social stability and economic health beyond the basic number of jobs available. The top 20% of households earned nearly half of all income (49.98%) in 2023, while the top 8% earned over a quarter (28.5%).

Lower-wage jobs are expected to expand as the service sector grows. By 2026, America is predicted to have more at-home caregivers than secretaries. Some experts, however, have a dim view of the jobs being created. David Blanchflower of Dartmouth College published a book in 2019, *Not Working: Where Have All the Good Jobs Gone?*, in which he argued that many workers are underemployed or have simply given up trying to find a well-paying job. He notes wage growth had not returned to pre-recession levels despite rosy employment indicators, and general prosperity has not returned since the crash of 2008. He said that the plight of the underemployed is contributing to widespread despair, a worsening drug epidemic, and the unchecked rise of right-wing populism.

Poverty in the US isn't the desperate condition it has been at times in the past. The Heritage Foundation published a controversial report in 2011: "Air Conditioning, Cable TV, and an Xbox: What is Poverty in the United States Today." Perhaps the message is that the deep frustration of those at the bottom of the income ladder is due to little prospect of climbing higher rather than basic economic insecurity. Today, the ability of high earners to get better treatment such as shorter lines at airports or better seats at sports events by paying more is very visible to those who can't afford to buy that better treatment. It's certainly not the same as not having enough to eat, but contributes to "the age of grievance," to borrow the title of Frank Bruni's 2024 book.

A June 20, 2025, article in the *New York Times*, "We need to make America Grateful Again," noted that Americans have it pretty good compared to other countries. It said that over the past half-century, child mortality has fallen by two-thirds in the United States, medical advances have made lives longer and more comfortable, education rates have soared, and material comforts like air-conditioning, plumbing and Internet access abound.

The article suggested that a key factor in Americans apparently being angry was "envy," the perception that others were doing better. That perception was driven largely by social media where people advertise their vacations to exotic locations, their happy family, their new house, their promotion, etc.

"An excellent civic education will cultivate reflective patriotism," Danielle Allen wrote in a 2023 essay for the *Washington Post* with the headline: "We hit rock bottom on civics education. Can we turn it around?" Many conservatives argue that schools overly teach shame at our racial history without a balance. Frank Bruni's *The Age of Grievance* said, "Grievance thrives on the idea that there's little good in and around us. But that's no less a myth than the insistence that the United States is a colorblind, classless utopia of unfettered opportunity."

Whatever the cause, it is clear that a large part of the population is focusing on their grievances rather than any potential positives we enjoy. This view minimizes the role of actual differences in income growth, which obviously play a major role as well.

What role does increasing computer power play in the inequality problem, if any? In one sense, it democratizes the information economy. Many information services, such as web search, are free if one has a PC or mobile phone. Similarly, digital assistants are free once you have a device that supports them. Streaming video has made it affordable to watch a movie every night if you wish, or you can binge-watch an old TV series. You can subscribe at a modest rate to a service that will play almost any song for you, e.g. by having an Amazon Echo speaker and an Amazon Prime subscription. Warehouses and delivery systems have become so efficient that something ordered online is often delivered within one or two days. There is a cost to basic service—wired or wireless connection to the Internet—but this only appears to be an issue in cases of extreme poverty. About 85% of the adult population in India, which grapples with high poverty rates, have access to a mobile phone. As computer power drops in cost, these services should expand and costs decline.

The impact of computer power on lowering some costs and increasing some services, however, will not help the core inequality of the distribution of wealth. Where it might help is by reducing the effect of differing education on earnings. A major determinant of income is a college degree.

This book has argued that a digital assistant that can communicate using human language can in effect make us smarter. In addition to providing an individual with an immediate source of specialized information, the assistant can reduce the need to have a college degree by being specialized to a job. As previously noted, warehouse stock picking by a worker to fulfill an order has been made more efficient and reduced errors by using a speech assistant for workers, reducing training requirements and making it possible to pay workers more. Such assistants are so effective that they are dominant in warehouses today.

As digital assistant technology becomes easier to implement, building such specialized assistants will become easier. For example, Amazon Web Services offers Amazon Q, which the company says "can be used as an intelligent virtual assistant across all business functions and by employees of all levels and capabilities. By simply chatting with Amazon Q in natural language, users can get help with their digital tasks." In a particular case, Deriv, one of the world's largest online brokers, faced challenges accessing vast amounts of data spread across various platforms. It adopted Amazon Q Business to retrieve and process data from multiple sources. Deriv reduced the time spent onboarding new hires by 45%, minimized recruiting time by 50%, and streamlined campaign management for marketing teams.

In another example of tools, Oracle's Digital Assistant delivers a complete AI platform to create conversational experiences for business applications through text and voice interfaces. Developers use the software to assemble digital assistants from one or more 'skills.' Skills can be based on prebuilt skills provided by Oracle or third parties, custom-developed, or based on one of the many skill templates available. Digital Assistant routes the user's request to the most appropriate skill to satisfy the user's request. Skills combine a multilingual NLP deep learning engine, a dialogue flow engine, and integration components to connect to back-end enterprise systems.

The value of a college degree seems to be reduced. A June 2025 article in *The Economist*, "Young Graduates Are Facing an Employment Crisis," notes that while the overall national unemployment rate remained around 4%, for new college graduates looking for work, it was 6.6% over the 12 months ending in May 2025. That is about the highest level in a decade—excluding the pandemic unemployment spike—and up from 6% for the 12-month period a year earlier.

This might be from AI taking over more jobs that previously required a degree. That would in a negative way reduce inequality. However, new job seekers always have a higher unemployment rate compared to workers with job experience, so the relatively small change may simply reflect the economic uncertainty in mid-2025.

A longer-term solution to inequality is improving education. In low- and middle-income countries, an estimated 70% of ten-year-olds cannot read a simple story in any language. In sub-Saharan Africa, the figure is closer to 90%. An AI assistant could allow an individual with a smartphone to learn at their pace.

Simply spending more money on conventional education doesn't always work. In Chicago total spending on schools passed $34,000 per pupil last year, compared with a national average of $20,000, but scores in reading and mathematics plunged over the past decade.

A working paper published in May 2025 by the World Bank suggested that AI may offer a partial solution. The study, reported by *The Economist*, followed 422 secondary-school students in Nigeria who took part in 12 90-minute after-school sessions over six weeks. Pairs of pupils, supported by a teacher, interacted with Microsoft Copilot, a chatbot based on GPT-4, to improve their English grammar, vocabulary and writing skills.

By the end of the six weeks the children in the AI 'treatment' group had made progress equivalent to nearly two years' worth of their regular schooling, according to Martín De Simone, who led the study. Overall, the AI group's test scores were about 10% higher than in the control group. In end-of-year exams—which covered topics beyond the chatbot's material—they still did better than their peers. (The final tests were done with pen and paper; the results reflected the children's learning, not their use of the tool.)

This experiment indicates the potential for helping students in any country who may have missed part of their schooling due to illness or who simply fall behind because of economic disadvantage that affects their learning. Parents or governments can encourage the use of educational software to help them catch up.

In the US, Duolingo, for example, offers software that helps students where English is a second language improve their English. Apex Learning provides virtual school options for grades 6-12. Amplify offers a variety of digital products and services for K-12 education, including core curriculum, assessments, and intervention programs. BrainPOP provides animated educational content for kids across various subjects. Newsela delivers news articles adapted for different reading levels, supporting literacy development. Zearn is a math learning platform designed to help students catch up and progress.

The American Federation of Teachers, the second-largest US teachers' union, said in July 2025 that it would start an AI training hub for educators with $23 million in funding from three leading chatbot makers: Microsoft, OpenAI and Anthropic. The union said it planned to open the National Academy for AI Instruction in New York City, starting with hands-on workshops for teachers on how to use AI tools for tasks like generating lesson plans.

The energy crisis

The rabid growth of computer centers supporting AI has led to questions of where the electric power to support them will come from. Companies such as Meta, Amazon, Microsoft, and Google have sent out a flurry of announcements related to nuclear energy. Some are about agreements to purchase power from existing plants, while others are about building additional plants.

MIT Technology Review quoted Michael Terrell, senior director of clean energy and carbon reduction at Google, in 2025: "There [are] a lot of advantages to nuclear...it's clean, firm, carbon-free, and can be sited just about anywhere." ('Firm' energy sources are those that provide constant power, as opposed to wind farms and solar.) But the publication cited a caution from Patrick White, former research director at the Nuclear Innovation Alliance, noting that there are needs on different timescales. Many of the tech companies will require large amounts of power in the next three to five years, but building new nuclear plants can take close to a decade.

It appears that the only way the world can both fight global warming and meet its energy needs in the long term is to use nuclear power. In the medium term, increasing green energy is only a partial solution. Instead, there is increasing use of energy created by burning oil, gas, and coal. The only long-term solution may be using the eventually available nuclear power in part to use carbon capture technology to take carbon dioxide out of the atmosphere. There may be a significant penalty during this transition, as extreme weather events already seem to be increasing.

However, one way this issue may be reduced would be due to a failure. I've argued that the huge sums being spent to grow ever-larger neural networks will not produce the expected results. They will provide very powerful computer centers needed for the machine learning process for the large neural nets and, to a lesser extent, running the models for use when they are complete. I discussed in the section on The Data Problem my skepticism that these bigger-is-better models would be successful. If I am correct, there will be a lot of computer power available for other things, perhaps more than needed, leading to a drop in cost of computer power. If it leads to much of this potential computer power being idle, it will also lead to lower energy requirements.

Motivating investment in increased nuclear power may be the greatest contribution of the bigger-is-better trend in AI.

Software as a 'non-rival' good

Paul Romer's paper, "Endogenous Technological Change," published in October 1990, went beyond the classical economic view that distinguished 'public' goods supplied by governments and 'private' goods supplied by the marketplace. Romer made a different distinction:

1. 'Rival' goods (goods with a physical embodiment that limits sharing, e.g. a candy bar, clothing, or a house). Once a candy bar is eaten, it can't be eaten again; only one person at a time can wear an item of clothing.
2. 'Non-rival' goods, such as knowledge that can be stored in a form such as computer bits, allow in principle unlimited sharing. You can send a copy of a digital photo to a friend and still have it. Non-rival goods, such as a digital version of a hit song, aren't necessarily free; Romer called them "partially excludable" in that an artificial price unrelated to the cost of production could be set by an "owner."

This subject has deep implications for economics. Much of classical economics has been defined by the consideration of rival goods. The balance of 'supply and demand' in classical economics usually assumes a limited supply. 'Scarcity' is sometimes even used in the definition of economics. For example, in the introduction to his book *Basic Economics* (4th edition, 2010), Thomas Sowell says:

Without scarcity, there is no need to economize—and therefore no economics. A distinguished British economist named Lionel Robbins gave the classic definition of economics: "Economics is the study of the use of scarce resources which have alternative uses."

The classical economists treated technology and similar factors that couldn't easily be put in their models as 'exogenous,' that is, a factor outside of economics that could impact, but was not part of, economics. In the title of his paper, Romer emphasizes it is 'endogenous,' something that must be dealt with in economic models. Software is a non-rival good, a form of knowledge representation that can be replicated as much as one wants. The fact that it has an increasing role in the economy reduces the importance of scarcity in economic models. Knowledge, one of the most important aspects of our economy, can today be easily shared and distributed.

David Warsh, in his 2007 book *Knowledge and the Wealth of Nations*, emphasized the impact eloquently, beginning with a familiar saying: "Give a man a fish, and you feed him for a day. Teach a man how to fish, and you feed him for a lifetime. To which it now must be added, invent a better method of fishing, or of farming fish, selling fish, changing fish (through genetic engineering), or preventing overfishing in the sea, and you feed a great many people, because these methods can be copied virtually without cost and spread around the world."

TV shows and movies are not as fundamental to existence as food, but are an important part of modern society and economics nevertheless. When one watches a TV show today, it doesn't disappear like fish when eaten. It still exists for someone else to watch as a digital asset. When one goes to a movie, the theater can be full or nearly empty, and the cost of projecting the movie doesn't change. When you watch a movie at home, there isn't even an incremental cost of presenting it (other than electricity), since you are using an existing TV set, PC, or mobile device. The movie is only scarce in the sense that the company holding the copyright controls access to it. Amazon Prime and Netflix allow watching thousands of movies for just one membership price, although Amazon charges a modest fee for some classic movies.

One of the most obvious areas of impact of non-rival goods is in books, music, and video. The ease of digital publishing is opening the way for creative talent that previously faced high hurdles to establish an audience. Writers can create a digital book by simply using an application on a web site to upload a text file and make it available through web-based outlets. Today, books can even be printed as orders are received, as opposed to long print runs that assume a large volume of sales. And, while it still requires substantial capital to make a full movie or TV series, anyone with talent can get at least some recognition by uploading a short original video to YouTube or TikTok.

I have argued that the human-computer connection can be a new base for rapid innovation and economic growth, driven by innovation in technologies such as speech recognition, natural language understanding, knowledge representation, and the digital assistant model, with these resources being almost continually available due to mobile devices. Because software is a non-rival good, its impact can grow very quickly. Unlike an innovation like railroads or the Internet, it doesn't require huge capital expenditures and a long time to have a major effect.

The bridge between humans and computers was limited in the past, and we are just seeing the beginning of its full construction. The wide embrace of that connection by consumers and businesses can be a new innovation that drives economic growth.

Will there be too much computer power?

Part of the rapid acceleration of growth in computer power is targeted at supporting generative AI, as noted previously. Generative AI is certainly impressive, but its very nature as a statistical process means it is difficult to incorporate into critical company processes. Statistical analysis summarizes the implications of datasets, and thus *approximates* what that data says. Companies that aren't careful how they use the technology internally may find themselves making costly mistakes. In customer-facing applications, errors in only a few cases could harm customers, e.g. saying their symptoms don't require going to an emergency room when they do (to take an extreme case). I suspect there will be a few lawsuits claiming harm from a generative AI suggestion, which certainly will cool adoption in customer-facing applications. (It will only take a few, even if they are frivolous.)

Thus, companies that are investing in larger computer centers to build such models may be disappointed in the results of that investment. However, once built, the centers provide computer power scaled for the intensive task of machine learning for a very large neural network. Even if the bigger networks do give some benefits, the need for computer power drops dramatically once the models are *used*, as opposed to being *built*. In any case, there will be a huge jump in computer power available. The most likely result is that some companies will turn to providing that excess computer power as a service, similar to Amazon Web Services (AWS) and Microsoft Azure. This could cause a major drop in the price of computer power. It's a bit like oil: when too much oil is produced, the price drops.

This could cause a problem for companies competing in this market. In the first quarter of 2025, revenues of Amazon Web Services rose 17%. AWS is one of Amazon's strongest revenue segments, generating over $115 billion in 2024 net sales, up from $105 billion in 2023. AWS generated $39.8 billion in operating income in 2024, a substantial portion of Amazon's overall operating income.

In the third quarter of financial year 2025, Microsoft Azure revenue growth stood at 33%. Microsoft's Intelligent Cloud segment, which includes Azure and other cloud services, contributed $26.8 billion to Microsoft's total revenue of $70.1 billion in the third quarter of fiscal year 2025.

Google Cloud is much smaller but growing. In Q1 2025, Google Cloud was 13.6% of the Google conglomerate's overall business and 7.1% of its operating income.

A major drop in the earnings generated by these cloud services could be significant, even for these giants. It will take a while, perhaps a few years, for generative AI computer centers to be completed, realize they have excess capacity, and enter the computer power market. The current players are investing huge amounts in growing capacity themselves. For example, Amazon reported capital expenditures of $26.3 billion during the final three months of 2024, primarily on infrastructure to support AI and the cloud, according to CEO Andy Jassy during an earnings call. The company expected to sustain that level of quarterly spending through the end of the year, exceeding $100 billion.

Of course, the lower cost of computer power can make many uses of that power more practical, as we have seen with AI. My candidate for absorbing that power is the rapid expansion of talk-or-text digital assistants, as I argued in the Talking to Computers chapter. The digital assistants will use increasing computer power to do speech-to-text conversion, natural language processing (often using neural networks), and LLMs to provide concise answers when possible. They will thus take advantage of this cheaper computer power.

To the degree that the assistants provide an alternative to classical web search, they could be a major source of advertising revenue—a further motivation driving their growth. The equivalent to a Google search producing a list of websites with sponsored sites listed near the top is a digital assistant giving direct conversational access to a company through the assistant. In the long run, this access may reduce the use of customer service telephone lines.

The digital assistants already have some ability to connect to other companies' apps. The idea is to be able to say or type something like "Please ask BigBank for my account balance." For example, Apple's "App Intents" framework can help a company expose their mobile app's functionality to Siri. The intents can take people directly deep within an app with Apple's "URL Representable Entities."

Google Assistant 'actions' are functionalities that extend the capabilities of the assistant beyond basic commands like setting timers or playing music. App Actions enable a company's users to say simple voice commands to quickly access an Android app's functionality. With Google Assistant's intent mapping and natural language understanding (NLU), developers can add a layer of voice commands and users can jump to the activities in an app where engagement matters most. Google gives the example of "Hey Google, send $28 to Joseph on Pay-Pal."

If the digital assistants increasingly perform the equivalent of web searches, these capabilities can be monetized. Some are partially monetized currently, e.g. Google's Conversational Agents, which help create chat conversations.

Companies will also be expected to allow their websites, mobile apps, and customer service lines to interact conversationally directly with the user using the digital assistant model rather than a click-here and then click-there model. The click model is similar to the touch-tone model for customer service ("Press 1 for our address and hours and 2 for everything else" followed by what seems like an endless number of options requiring pressing keys). Speech recognition is slowly replacing that frustrating experience with something like "Please state why you are calling."

Eventually a company will require a company digital assistant as much as they require a web site. That company digital assistant will provide the functionality to connect through the general digital assistants as well as directly with customers.

Bank of America's Erica is an example of a company digital assistant. You can talk or type to Erica. Some of Erica's capabilities, according to the company, are:

- *Get account information:* Access your routing and account numbers, find companies that have your card information stored, and access your security meter to help you protect your accounts;
- *Search Transactions:* Locate transactions including groceries, utility payments, transportation expenses and more;

- *Manage your cards:* Temporarily lock or unlock your misplaced debit card, replace a lost or stolen card, and check balances;
- *Live chat:* Start a live chat with a specialist;
- *FICO® Score:* Provides your FICO® credit score and notifies you about important credit score changes as they occur;
- *Monitor recurring charges:* View recurring charges, receive notifications when a recurring charge or membership cost increases, and more;
- *Track refunds:* Keep track of your refunds so you know when money from a return is available in your account;
- *Track spending by category:* View monthly spending by category (e.g. Utilities, Groceries, Transportation, Shopping);
- *Bank of America Preferred Rewards®:* Notify you when you're eligible for Preferred Rewards, explain the benefits and help you enroll;
- *Make Merrill investing easier:* Access quotes, track performance, place trades and connect you with a Merrill advisor; and
- *Bill Reminders:* Remind you of all your upcoming e-bills and help schedule payments.

There will be some turbulence as the large companies compete to be your digital assistant. As companies move R&D expenditures from generative AI to digital assistants, I expect a lot of commentary on how far beyond the Turing Test of whether a digital assistant can simulate a human we have gone.

12

War: Declared or Otherwise

The Russia-Ukraine war has exposed a major change in the way future wars will be fought. Some of that change involves adding intelligence to weapons such as missiles and unmanned aerial vehicles (UAVs) such as drones. Another possibility is cyberwarfare, attacking infrastructure such as power stations; cyberwarfare can be denied, allowing undeclared war on both economic and military targets at a distance.

Computer Power in War

AI and computer analysis in general is increasingly integrated into military operations across several key areas:

- *Surveillance and Intelligence*: AI processes vast amounts of data from satellites, drones, and sensors to identify targets, track movements, and analyze patterns. Machine learning algorithms can detect changes in terrain, identify vehicles or personnel, and flag unusual activities that might indicate threats.
- *Autonomous and Semi-Autonomous Weapons:* Some weapons systems use AI for navigation and target identification. These range from defensive systems that intercept incoming missiles to loitering munitions that can identify and engage targets. The level of human oversight varies significantly.

- *Cyber Operations:* AI enhances both offensive and defensive cyber capabilities. It can automate the detection of vulnerabilities, respond to cyber-attacks more quickly than humans, and potentially launch sophisticated cyber operations at machine speed.
- *Logistics and Support:* Military computer systems optimize supply chains, predict maintenance needs for equipment, and manage resource allocation. This includes route planning for convoys and coordinating complex operations across multiple units.
- *Decision Support:* AI assists commanders by processing intelligence, modeling scenarios, and suggesting tactical options. It can rapidly analyze battlefield conditions and provide recommendations, though humans typically retain final decision-making authority.
- *Electronic Warfare:* AI helps manage the electromagnetic spectrum, jamming enemy communications while protecting friendly signals. It can adapt to countermeasures and optimize electronic warfare strategies in realtime.

In June 2025, *The Economist* posted an article, "How AI is changing warfare," with the subtitle "An AI-assisted general staff may be more important than killer robots." It highlighted how AI integrating information from multiple sources such as satellites, drones, and soldiers on the ground could give effective advice on strategy to military leaders. Simulations showed the effectiveness of such systems.

Computer intelligence in drones is advancing. Both Russia and Ukraine have been rushing to develop software to make drones capable of navigating to and homing in on a target autonomously using chips in the drones, even if jamming disrupts the link between pilot and drone. Videos of drone strikes in Ukraine increasingly show 'bounding boxes' appearing around objects, suggesting that the drone is identifying and locking on to a target.

In spring 2025, Ukraine announced plans to deploy 15,000 ground robots—unmanned ground vehicles (UGVs). According to *The Econo-*

mist, some analysts predicted that the face of the battlefield will rapidly change in the summer of 2025, likening the proliferation of UGVs to the explosion in aerial-drone manufacturing in 2023. One Ukrainian manufacturer said, "We don't have the men to counteract Russia's meat-wave. So we'll send our own zombies against theirs."

But the utility of AI in more prosaic applications is equally effective. A recent study by think-tank the RAND Corporation found that AI, by predicting when maintenance would be needed on A-10C warplanes, could save America's air force $25 million a month by avoiding breakdowns and overstocking of parts. Logistics was another promising area. The *Economist* article said the US Army was using algorithms to predict when Ukrainian howitzers would need new barrels, for instance. The army is also using a model trained on 140,000 personnel files to help score soldiers for promotion.

The US began Project Maven in 2017 to deal with the deluge of photos and videos taken by drones in Afghanistan and Iraq. Maven is "already producing large volumes of computer-vision detections for warfighter requirements," noted the director of the National Geospatial-Intelligence Agency, which runs the project, in May 2025. The stated aim was for Maven "to meet or exceed human detection, classification, and tracking performance." *The Economist*'s tracker of war-related fires in Ukraine is based on machine-learning, is entirely automated, and operates at a scale that journalists could not match, according to the publication. It had already detected 93,000 probable war-related blazes in June 2025.

The US Air Force recently asked the RAND Corporation to assess whether AI tools could provide options to a "space warfighter" dealing with an incoming threat to a satellite. The conclusion was that AI could indeed recommend "high-quality" responses.

Tamir Hayman, a general who led Israeli military intelligence until 2021, pointed to two big breakthroughs. He said the "fundamental leap," eight or nine years ago, was in speech recognition software that enabled voice intercepts to be searched for keywords. The other was in computer vision. Project Spotter, in Britain's defense ministry, is already using neural networks for the "automated detection and identification of objects" in satellite images, allowing places to be "automatically monitored 24/7 for changes in activity." *The Economist* reported that, as of February 2025, a private company had labelled 25,000 objects to train the model.

The Economist quoted General Hayman: "If...machines produce a lower false positive and false negative rate than humans, particularly under pressure, it would be unethical not to delegate authority. We did various kinds of tests where we compared the capabilities and the achievements of the machine and compared to that of the human. Most tests reveal that the machine is far, far, far more accurate...in most cases it's no comparison."

There are questions about how much of the theory has truly penetrated current defense practices. *The Economist* quoted Sir Chris Deverell, a retired British general: "The irony here is that we talk as if AI is everywhere in defense, when it is almost nowhere. The penetration of AI in the UK Ministry of Defence is almost zero...There is a lot of innovation theatre." And the Pentagon spends less than 1% of its budget on software.

The use of AI in warfare raises significant questions about accountability, escalation risks, and the role of human judgment in life-and-death decisions. The accurate identification of military targets can, on the other hand, reduce civilian casualties.

Cyberwar

Cyberwar is an attack where digital systems are compromised by malware delivered over the Internet surreptitiously or simply by overwhelming a web site by heavy traffic—a 'denial-of-service' attack. Cyberattacks can be used to bring down a software system or even damage physical infrastructure by attacking computerized control systems. Many cyberattacks are delivered by individuals or groups not associated with governments, in some cases as 'ransomware' with a demand for money.

When sponsored by governments, cyberattacks can be directed at stealing classified information on weapon systems, delivering propaganda, or shutting down critical infrastructure. The use of cyberattacks in elections to steal data such as a politician's emails and to defame or promote a particular candidate have been publicized.

Russian interference efforts in the 2024 US presidential election were extensive and involved multiple tactics. Russia spread disinformation before the 2024 election to damage Joe Biden and Democrats, boost candidates supporting isolationism, and undercut support for Ukraine aid and NATO. Their efforts included disinformation campaigns, hoax bomb threats linked to Russia targeting polling places in battleground states (with at least two polling sites in Georgia briefly evacuated), and cyber influence operations and fake videos that received millions of views. The interference efforts were part of a broader pattern involving Russia, Iran, and China, with each country employing different tactics to influence American voters and undermine confidence in the electoral process.

Malware—attack software—has been used specifically as a tool of war. In 2010, Stuxnet, developed by the US National Security Agency, perhaps with the aid of Israeli intelligence, targeted the software of physical controllers and caused centrifuges used in the enrichment of uranium at a facility in Iran to self-destruct. Malware called BlackEnergy was used by Russian hackers to launch an attack in December 2015 on several Ukrainian power companies. The malware was used to gather intelligence about the power companies' systems and to steal login credentials from employees; it was then used to trigger blackouts.

Other malware developed by the Russians, Industroyer, was used to mount an attack on a part of Ukraine's electrical grid in December 2016. The code was used to strike an electrical transmission substation in Kiev, blacking out part of the city for a short time. In June 2017, the Notpetya virus was spread in the Ukraine through a software update for a popular Ukrainian accounting software package used by about 80% of Ukraine's businesses. The software masqueraded as a ransomware virus, but in fact destroyed data to incapacitate computer systems. The attack was widely regarded as part of Russia's continuing cyberattack on Ukraine.

In a talk in November 2018, Microsoft President Brad Smith, reacting in part to the Russian Notpetya attack, drew a parallel between the run-up to the First World War and the burgeoning cyberwar arms race. "I'm not here to say the next world war is imminent, but I am here to say that there are lessons from a century ago we can learn and apply, that we need to apply, to our own future," Smith said. "When we are talking about cyberspace, fundamentally we are talking about space that is private property, we're talking about datacenters and undersea cables and laptops and phones and devices and services that we create. Like it or not, and I don't think we should like it, the reality is inescapable; we have become the battlefield."

An article in the *Wall Street Journal* on June 29, 2025, described cyberwarfare conducted by Isreal against Iran as part of its attack on Iran's nuclear capabilities. Israeli authorities, and a pro-Israeli hacking group called Predatory Sparrow, targeted financial organizations that Iranians use to move money and sidestep the US-led economic blockade. Predatory Sparrow, which operates anonymously and posts updates of its activities on X, said it crippled Iran's state-owned Bank Sepah, which services Iran's armed forces and helps them pay suppliers abroad, knocking out its online banking services and cash machines. Iranian state media acknowledged the damage. The group also breached Nobitex, Iran's largest cryptocurrency exchange, popular with locals for transferring money overseas. The hackers extracted about $100 million in funds and forced the platform to shut down, according to the exchange. Predatory Sparrow didn't say if it was acting on behalf of Israeli authorities.

Malware called Triton is believed to be the creation of Russian state-sponsored hackers. In March 2019, journalist Martin Giles called Triton "the world's most murderous malware" because it could disable safety systems designed to prevent catastrophic industrial accidents. The malware could, at the hackers' command, prevent physical controllers from preventing disasters. It could disable emergency shut-off valves or pressure-release valves that were designed to work automatically when they detected dangerous conditions. Triton was discovered at a petrochemical plant in Saudi Arabia when a flaw in the code triggered several unintended shutdowns.

On March 7, 2017, WikiLeaks published 8,761 documents allegedly stolen from the CIA of alleged spying operations and hacking tools. These included iOS and Android vulnerabilities, bugs in Windows, and the ability to turn some smart TVs into listening devices. This obviously harmed CIA operations. It also raised the question of whether the CIA should have revealed those vulnerabilities itself to avoid their being used by criminals or other countries, as opposed to exploiting them for its own purposes.

Our most critical systems in part suffer from old software and poor procedures to protect it. A security audit of the US ballistic missile system released in December 2018 by the US Department of Defense Inspector General found that several ballistic missile facilities lacked data encryption, antivirus programs, and multifactor authentication mechanisms, and had 28-year-old unpatched vulnerabilities.

The US has resources for cyberattacks, and we of course don't know all the activities of US intelligence agencies. The *Washington Post* reported in February 2019 that the US Cyber Command targeted the St. Petersburg-based Internet Research Agency with a cyberattack in late 2018 that knocked the organization offline during the US midterm elections, potentially preventing a last-minute flood of disinformation designed to affect the election's results or turnout.

There is potentially a lesson from nuclear weapons. It would be almost impossible to prevent massive destruction in a full nuclear attack; what deters that frightening possibility is the similar destruction a counterattack would deliver. Nations must unfortunately develop the ability to counterattack with cyberweapons, a bit like the mutually assured destruction of the nuclear age.

Hackers have even attacked companies that develop anti-malware software. A group of Russian hackers claimed in May 2019 to have infiltrated the networks of three US-based malware detection software providers and stolen the source code for their software. The hackers claimed to have stolen 30 terabytes of data and were demanding a ransom to give it back and to not name the companies attacked.

Valery Gerasimov, a Russian general, elucidated in a military journal in 2011 what has become known as the Gerasimov Doctrine: "The role of nonmilitary means of achieving political and strategic goals has grown. In many cases, they have exceeded the power of force of weapons in their effectiveness." The doctrine was formally incorporated into Russian military strategy in 2014. About 75 Russian research institutions were devoted to the study and weaponization of information, coordinated by the Federal Security Service (the successor of the KGB).

The Russian government claimed that, even if cyberattacks against Ukraine originated in Russia, they were the actions of isolated, patriotic individuals who wanted to promote Russian interests. This attempt at denying the government's role is a major characteristic of cyberwarfare; it can cloud the motivation for a counterattack. Counterattacks—even if unannounced—are one way to discourage such activities. The Ukrainian Cyber Alliance, a group of Ukrainian hackers that works to fix security breaches, took credit for several attacks carried out in late 2016 against the Russian government and government agencies.

The *New York Times* reported in June 2019 that the US was stepping up digital incursions into Russia's electric power grid in a warning to President Putin. The article said that the Trump administration was using new authority to deploy cybertools more aggressively. US government officials indicated that there was a previously unreported deployment of US computer code inside Russia's power grid and other targets. The initiatives were more aggressive than probes in the past and capable of damaging the targeted systems. Presumably, the publicity about the cyberattacks was designed to warn the Russians of consequences if they activated any similar malware already installed in the US. In a public appearance in June 2019, National Security Adviser John Bolton said the US was now taking a broader view of potential digital targets as part of an effort "to say to Russia, or anybody else that's engaged in cyberoperations against us, 'You will pay a price.'"

New authority to do so was granted separately by the White House and Congress in 2018 to the United States Cyber Command. In the military appropriations bill in summer 2018, Congress approved the routine conduct of "clandestine military activity" in cyberspace to "deter, safeguard or defend against attacks or malicious cyberactivities against the United States."

In July 2018, Department of Homeland Security Secretary Kirstjen Nielsen said cyberattacks pose a greater threat to the US than physical ones. In June 2018, a group of presidential advisers said the country needs to prepare for a "catastrophic power outage" possibly caused by a cyberattack. The National Infrastructure Advisory Council, mostly current or former chief executives of companies engaged in critical industries, said resources need to be stockpiled in community enclaves to prevent mass migrations of desperate people in the event of a long power loss.

In July 2018, the US Department of Homeland Security (DHS) reported that a group of Russian hackers (called 'Dragonfly'), apparently supported by the Russian government, had installed software in the control systems of US electric utilities. The hackers got in relatively easily by penetrating the networks of outside vendors supporting the utilities. "They got to the point where they could have thrown switches" and disrupted power flows, said Jonathan Homer, chief of industrial-control-system analysis for the DHS, as quoted in a *Wall Street Journal* article. DHS said hundreds of utilities were affected by Dragonfly. Researchers attending the CyberwarCon forum in November 2018 in Washington, D.C., also warned that Russian cyber attackers are targeting the US electrical grid, searching for vulnerabilities to intrude on electricity generation and transmission systems.

The DHS released in April 2019 a list of "national critical functions," defined as "the functions of government and the private sector so vital to the United States that their disruption, corruption, or dysfunction would have a debilitating effect on security, national economic security, national public health or safety, or any combination thereof." The list of over 50 items included, for example, fuel refining, the provision of information technology products and services, protecting the water supply, electricity distribution, air transportation, conducting elections, medical care, and Internet routing and access. Addressing any one of these threats effectively is a huge task. The objective was to use the list to identify the biggest risks.

Chris Krebs, director of the DHS's Cybersecurity and Infrastructure Security Agency, told the *Washington Post* at the time: "If everything's a priority, then nothing's a priority. This allows us to really drill down into those things we need to care about." On April 30, 2019, the DHS issued a "binding operational directive" that required agencies to remediate "critical" vulnerabilities identified by the Cybersecurity and Infrastructure Security Agency (CISA) within 15 days of detection, a reduction from 30 days. The directive also established a new lower-priority category, "high" vulnerabilities, which must be remediated within 30 days.

Russia isn't the only country active in penetrating US facilities and companies. China has a deep commitment to hacking and other techniques but has emphasized economic targets more than Russia. China's Ministry of State Security and the People's Liberation Army are believed to have stolen the design details of many pieces of American military hardware, from fighter jets to robots, a claim that the Chinese have persistently denied. In 2012, National Security Agency Director Keith Alexander called it the "greatest transfer of wealth in history." President Obama pressed the issue of cyberthefts in his first meeting with President Xi in 2013, only to be met with more denials.

Regarding potential cyberattacks on the US, the US Defense Department does not appear to have made cybersecurity a priority in the past. A nearly one-year audit of cybersecurity, completed in November 2018 at a cost of over $400 million, suggested not enough was being done. "We failed the audit. But we never expected to pass it," Deputy Secretary of Defense Patrick Shanahan told reporters at its completion. He said the audit revealed many issues, including inventory inaccuracy and cases of not complying with cybersecurity discipline. The most cited problems were related to the information technology security of the Defense Department's business systems. Auditors found the department's "financial and business management systems and processes do not provide reliable, timely, nor accurate information." The report also states that the department's IT has "systemic shortfalls in implementing cybersecurity measures to guard the data protection environment" and "issues exist in policy compliance with cybersecurity measures, oversight, and accountability."

The government also published the National Cyber Strategy for United States of America in September 2018. The introduction by President Donald Trump indicated that National Cyber Strategy is the US's "first fully articulated cyber strategy in 15 years." The 26-page document includes multi-agency general objectives, such as securing critical infrastructure, combating cybercrime, fostering a stronger cybersecurity workforce, promoting responsible behavior between nation states, and preventing malicious "information campaigns." Such goals remain as words, however admirable, unless they are translated into specific funded activities.

In 2018, the US government made cybersecurity more of a priority with two initiatives: The Small Business Cybersecurity Act was signed into law. The bill amended the National Institute of Standards and Technology (NIST) Act to require NIST to include small businesses in its initiatives. NIST facilitates and supports the development of voluntary, consensus-based, industry-led guidelines and procedures to cost-effectively reduce cyber risks to critical infrastructure. NIST was to disseminate and publish on its website "standard and method resources that small business may use voluntarily to help identify, assess, manage, and reduce their cybersecurity risks."

The Trump administration's indiscriminate firing of federal employees at the start of his second term included NIST. Several media outlets reported that hundreds of employees at NIST would lose their jobs. Axios, citing anonymous sources "familiar with the matter," reported that the Gaithersburg, Maryland-headquartered agency plans to fire or lay off almost 500 people. The layoffs reportedly include many of the staffers behind the US AI Safety Institute (AISI), which NIST created in 2023 to further the AI safety and security goals in then-President Joe Biden's AI executive order.

Cyber threats are currently evolving rapidly as adversaries become more sophisticated and the number of connected devices worldwide continues to rise. New research reveals that more than 30,000 vulnerabilities were disclosed in 2024, a 17% increase. Gartner estimates global IT spending grew at an 8% rate in 2024, reaching $5.1 trillion, with 80% of CIOs increasing their cybersecurity budgets.

According to a July issue of the *Washington Post*, undeterred by recent indictments alleging widespread cyberespionage against American agencies, journalists and infrastructure targets, Chinese hackers were hitting a wider range of targets and battling harder to stay inside once detected, seven current and former US officials said in interviews. Hacks from suspected Chinese government actors detected by security firm CrowdStrike more than doubled from 2023, rising to 330 in 2024, and continued to climb as the new administration took over, CrowdStrike said.

Attacks are so frequent that they seldom make the news. A few examples:

- In July 2025, Qantas said it had been hit by a massive cyber-attack, exposing the personal data of around 6 million customers.

The Australian airline said it took "immediate steps" to tackle the breach but expected a "significant" portion of data to be stolen.

- In 2025, a Qilin ransomware attack on the UK's NHS resulted in a confirmed patient death, highlighting the life-threatening impact of healthcare cyberattacks.
- In 2025, a hacker stole a record $1.46 billion from Bybit ETH cold wallet.
- Thousands of employees, customers and business partners of Japanese electronics manufacturer Casio had data stolen during a ransomware attack in October 2024. Casio said that 6,456 employees, 1,931 business partners and 91 customers were impacted by the ransomware incident.
- In February 2025, Lee Enterprises, one of the largest newspaper groups in the United States, said a cyberattack that hit its systems caused an outage and impacted its operations. Distribution of products, billing, collections, and vendor payments were all affected. Distribution of print publications across the portfolio of products experienced delays, and online operations were partially limited.
- In 2025, Australian in-vitro fertilization giant Genea confirmed hackers "accessed data" during a cyberattack. The incident disrupted patient services and potentially led to sensitive information being accessed.
- Blood-donation not-for-profit OneBlood confirmed that donors' personal information was stolen in a ransomware attack in the summer of 2024. OneBlood said that ransomware actors had encrypted its virtual machines, forcing the healthcare organization to fall back to using manual processes.
- A ransomware attack on a large healthcare network in Maryland in early 2025 forced officials to shut off IT systems and cancel some appointments as Frederick Health Medical Group warned that there will be delays in service as it contended with the cyberattack.

- In 2025, the International Civil Aviation Organization (ICAO), a part of the United Nations, confirmed a hack of its recruitment systems involving the compromise of more than 40,000 records containing personal information. The threat actor known as 'Natohub' on the hacking forum BreachForums 2 compromised 42,000 documents from ICAO containing personal data. According to Natohub, the ICAO personal records include full names, dates of birth, physical and email addresses, phone numbers, and details about the individuals' education history and employment.
- Football team Green Bay Packers said cybercriminals stole the credit card data of over 8,500 customers after hacking its official Pro Shop online retail store in a September 2024 breach.
- Medusind, a leading billing provider for healthcare organizations, notified hundreds of thousands of individuals of a data breach that exposed their personal and health information in December 2023.
- At least two US school districts (South Portland Public Schools in Maine and Rutherford County Schools) suffered cyberattacks over the Christmas and New Years holidays in 2024, continuing an annual trend of hackers targeting K-12 schools and colleges during periods when IT staffing was at its lowest. South Portland Public Schools in Maine said it was forced to take its network down, and Rutherford County Schools said that it had been dealing with "network and systems disruption."
- In early 2025, Chinese AI platform DeepSeek disabled registrations on its DeepSeek-V3 chat platform due to an ongoing "large-scale" cyberattack targeting its services.

The cybersecurity landscape in 2025 is characterized by increasingly sophisticated attacks leveraging AI, while organizations are responding with enhanced zero-trust architectures and AI-powered defense systems. Most known attacks are by hacker groups rather than countries.

It is likely that countries set up software that could be activated as a response to a major attack by another country—the mutually-assured-destruction model.

13

The Evolution of
Human-Computer Interaction

Human evolution has always involved extending the core capabilities of our bodies and minds with external inventions. Our clothing extends the limitations of our bodies to withstand extremes of the environment, and, although we may change our clothing every day, it is almost part of us. Our automobiles in effect give us motorized wheels that go beyond what we can do with our legs alone. Our telephones extend the reach of our voice. We meet with widely dispersed individuals on Zoom. We don't consider these extensions part of ourselves, but we certainly depend on them constantly. At work we use tools, ranging from hammers to PCs, to go beyond what our bodies and minds alone can do. We transfer knowledge through books and media much more efficiently and to more people than face-to-face conversation would allow.

Expanding our humanity through our inventions is nothing new. Some connections with technology do become part of our brain. We drive a car, ride a bicycle, read, or type without thinking about all the detailed actions and processing needed to make this happen. They become part of our autonomous nervous system, embedded in our synaptic connections between neurons. We certainly didn't get these skills though evolution, yet we can use them even if we ignore them for years (e.g. riding a bike).

If we view computers as partners, not competitors, they in effect change what it means to be human. Computer technology can be our constant partner in getting things done. We don't have a safe way to add neurons to our brain; a tighter connection to computer power is an alternative that could have nearly the same effect. Computer power is driving the next stage in human evolution.

UC Berkeley robotics professor Ken Goldberg, chief scientist, Ambi Robotics and Jacobi Robotics, has argued against fearing AI and for the idea of humans and machines working together, what he called "multiplicity." He argues that computers multiply human intelligence rather than diminish it. Most technology today involves digital processing, with varying degrees of software complexity. That technology has largely seemed external, part of society, but separate from an individual human. As computer technology evolves to include more of what we call 'intelligence' and its daily connection with us tightens and is increasingly always available, it comes closer to a change in what it means to be a human.

Whether you want to call digital assistants the next step in human evolution or simply technology continuing to have an increasing impact on living is largely a matter of definition. But let's consider how it will likely affect the children who grow up with that close computer connection.

Jane cries in her crib not long after birth. A device called a 'baby tender' on her crib recognizes the cry. It begins playing soothing music. It signals the mother's smartphone. She acknowledges the alert but must finish a task before coming. She speaks through the device to the baby: "I'm coming, sweetheart. Be there soon." The baby may continue to cry but is reassured through the device that her mother is coming. The baby is indirectly becoming accustomed to a digital companion.

As Jane gets older, she will have at least one toy with a built-in computer chip connected wirelessly to a central system in the home. The central system can access information through the Internet. A toy bear with a screen on its tummy may talk with her, maintaining a conversation. It can display pictures to enhance the conversation. It supports early-education features. The toy is in effect her first digital assistant, interacting with her conversationally and encouraging her to speak.

Jane gets her own child-oriented smartphone or a similar device with smartphone features such as a smartwatch at an early age, if only to be able to track the child's location and warn if the child goes outside a specified perimeter.

She converses with the digital assistant to hear information or call friends or family as she gets older. The assistant alerts her if a friend is calling, perhaps increasing rather than decreasing social interaction. She learns to read early because of a reading game that can speak words on the screen and correct mistakes when she reads out loud. Her access to facts and vocabulary is expanded even before she can read and write.

Parents will feel pressure to provide this tool to their child as they see other children benefiting from similar digital assistants. If Jane has a problem with learning, her digital assistant can help. Digital assistants can help level the economic playing field for those whose ability level or early family disadvantages make learning more difficult; a country could assure the availability of such assistants to all. The assistant can help with talents other than academic through introducing hobbies that leverage other aptitudes.

Jane's digital assistant is available through several devices over time. It becomes increasingly personalized as she uses it. Her first instinct if she needs to know something is to ask the assistant. The assistant attempts to give as concise an answer to questions as possible. It can engage in a dialog to clarify a request when it doesn't fully understand.

The assistant can also be proactive if it thinks it perceives a need based on either a current conversation, a calendar entry, an incoming message, senses a fall, notes that Jane is ignoring fitness activities, or detects a possible medical problem. As she gets older, Jane comes to depend on the assistant to label messages by priority and block otherwise overwhelming spam. As she moves into adulthood, Jane's assistant continues to be available on multiple devices, including speakers throughout her home.

When Jane is old enough to drive, the car becomes an extension of the digital assistant, most obviously giving driving directions and using sensors built into the car to warn of driving hazards. The assistant is supported by a large company, as we see today, competing to be her portal to the web and the ability to offer optional services for a fee. It is periodically improved with increasing accuracy and features.

Jane can pay for services matching her interests. Jane often buys items through her digital assistant, using it much like web search. Online sales outlets deliver her requested items to her quickly. The trend to buying online is augmented as the language user interface makes it easy to ask questions about a product before buying without calling a customer service line. When a screen is available, she can view product options. She can connect directly with companies providing products of interest using a conversational interface.

She controls connected devices in her home with voice commands. She may change the target temperature of a thermostat, turn on a light, or speak to someone ringing the doorbell.

The assistant also becomes technical support as computer applications become increasingly complex and constantly change through automated software updates. She can ask how to do certain things with a piece of software in both personal and work environments. If the digital assistant can't interact with the software directly through an application programming interface provided by the software developer, it can walk Jane through the steps to reach her goal. Digital assistants as interfaces to other software become almost a requirement as the number of features in software packages and web services grow and become more complex.

A child who grows up with such experiences would almost view the connection with computer intelligence as being part of them. Jane's personal assistant's help seems a natural part of being a modern human. If the assistant is hacked or malfunctions, it is almost like contracting a physical illness. (Even today, most people view losing a mobile phone as a catastrophe.) Until the assistant is cured, Jane loses an ability she takes for granted. A major and critical feature of the assistant functionality is thus backup of data and cybersecurity. Jane may also develop an emotional connection, much like the affection one has for a pet.

There is no technological hurdle to making this scenario a practical future. Everything described is available today with some limitations. Personal assistants such as Amazon's Alexa, Google Assistant, and Apple Siri are used extensively. These large companies are investing heavily in making them understand speech better and provide crisp answers to questions when possible. The generative AI models are increasingly becoming digital assistants, often allowing interaction by voice or text.

You will certainly see concerns expressed regarding young people increasingly depending on a digital assistant, but the trend will be hard to resist, particularly with many companies promoting it. Even if some parents hesitate to provide devices to their children, suppliers of the general personal assistants will be motivated competitively to ensure broad early adoption, potentially by including some early-education software as part of the software delivered with the device. The philosophy 'no child left behind' might even motivate a government program providing devices with digital assistants and educational software for low-income families. Such a plan must recognize that those devices today require an Internet connection with WiFi or similar connectivity or cellular service, both of which come with an expense to set up and typically monthly fees to maintain. Some innovation and perhaps support by technology giants will be required to make this connection to computer support universal and prevent it widening the inequality divide instead of reducing it.

There will, of course, be concern that digital assistants are reducing parental interaction with the child, damaging the child's emotional growth and social skills. The same arguments are made about television and game systems, but those concerns haven't prevented those systems from being available to most children. It will remain a parental responsibility to encourage direct social interaction. Some trends take a long time to develop, and digital assistants becoming part of being human may take some time. Nevertheless, the trend seems inevitable in the long term.

Partnering with computers in getting things done

Arguing that digital assistants will become part of being human might be considered an extreme position. Computers can become partners in helping us get things done without reaching that extreme. Many AI innovations take advantage of machine learning to learn a particular skill that can aid in specialized areas. For example, in enterprise applications they can guide the analysis of complex data by a human analyst to uncover implications of that data that might otherwise be missed.

One example is using deep learning to analyze a database of chest x-rays where a radiologist has identified likely tumors. The result of the analysis is an algorithm that can highlight suspicious areas in a new x-ray being examined by a radiologist. Speech recognition can allow dictation of a report by the radiologist. The technology is a partnership in that it doesn't attempt to replace a radiologist, only help the radiologist be more efficient and accurate. The final diagnosis and recommendation for further action is always done by a human, an example of the human-computer partnership.

This example is not speculation; most radiologists use a computer assistant today. In a recent review, "Redefining Radiology: A Review of Artificial Intelligence Integration in Medical Imaging," Reabal Najjar said, "AI, particularly its subset machine learning, is radically improving radiology, strengthening image analysis, and mitigating diagnostic errors. AI algorithms process and interpret data, performing tasks that emulate or even surpass human cognitive capabilities."

There are similar uses in other areas of healthcare. The Mass General Brigham healthcare system in Boston, which includes Brigham and Women's Hospital and Massachusetts General Hospital, had developed more than 50 algorithms for use in its clinical practice as of mid-2021, some of which had been FDA-cleared. One example is an algorithm using machine learning to help detect abdominal aortic aneurysms. It quickly identifies the presence or absence of an aortic aneurysm, a bulge or ballooning in the wall of the aorta, the main artery carrying blood from the heart. Mass General Brigham has made cleared algorithms generally available as a commercial product sold by Nuance Communications (now part of Microsoft).

In industry, employee turnover is an increasing problem. Training new employees in jobs such as production lines, warehouses, or repair can be difficult if they must learn on the job. Virtual reality, wearing eyewear that superimposes instructions as the employee views the task at hand and allows interaction by voice, is an increasingly effective approach to on-the-job training.

Another very different type of partnership is machine learning systems that physicist Mario Krenn built to help understand the mysteries of quantum theory. The software Krenn wrote could take an experimental setup and calculate the resulting output using the same building blocks that experimenters use to create and manipulate photons on an optical bench, a vibration control platform that is used to support systems used for optics-related experiments. These tools include lasers, nonlinear crystals, beam splitters, phase shifters, and holograms. The program searched through a huge number of possible configurations by randomly mixing and matching the building blocks and calculated what the result would be.

The result revealed some unexpected conclusions. The most interesting could be reviewed by physicists to see why the experimental setup might yield such results. The experiments provided important insights into the mysterious instantaneous action-at-a-distance called 'entanglement.' Again, this represents a partnership where computer analysis pointed to one of many experiments that, if tried randomly, would have taken a much longer time.

Often it isn't obvious computers are helping. When we drive a modern car, our steering wheel doesn't turn the wheels and stepping on the brake pedal doesn't press the brake against the wheel; these actions instruct microprocessors to pass on the command to digitally controlled actuators. This is so intrinsic to automobile operation that a shortage of chips has at times significantly slowed auto production lines.

Our partnership with computer technology will grow. Of course, digital assistants are the ultimate computer partner.

14

What Does It All Mean?

In this concluding chapter, I'll summarize a few key points I've covered in more detail in preceding chapters:

- Accelerating computer power is a force driving change independent of extremes such as artificial intelligence.
- 'Artificial intelligence' today is focused on one technology: machine learning using deep neural networks.
- Moving from today's effective AI to 'artificial general intelligence' or 'superintelligence' is a vague and questionable goal.
- Creating very large neural networks is limited by data availability; simply making the networks larger has limitations.
- Computer power growth today is accelerating far beyond the doubling every two years as predicted by Moore's Law.
- The rapid change in computer power will impact the overall economy.
- The dangers of AI are overstated but deserve consideration. Nevertheless, increasing computer power will drive change despite those concerns.
- The AI technology of 'talking to computers'—speech recognition—has reached an impressive level of performance and will continue to improve as it supports more dialog and more understanding of what we say or type.

- Always-available talk-or-text digital assistants are the next revolution in computing, driven in part by rapidly increasing computer power.

Computer power growth is accelerating

The exponential growth in computer power each year drives much of our experience with technology and its impact on society. The attention given to artificial intelligence misses the reality that the core technology behind today's AI, machine learning using deep neural nets, was not a result of some new invention, but simply computer power passing a threshold that made deep learning, a technology known for decades, economically feasible. That point suggests that we should look at the growth in computer power in general as a major factor driving technological and societal change. That focus allows us to ask what comes next after neural networks.

What is summarized here as growth in 'computer power' has multiple components:

- Moore's Law allows growth in the amount of computation possible *on a single chip* by putting more transistors on an integrated circuit.
- Beyond a single chip doing more, there are *more chips* as companies add servers to their computing centers and consumers buy devices such as PCs and smartphones that do computation at the 'edge' of the network. Every upgrade to a new PC or smartphone and every expansion of a company computer center adds to the total of computer power available.
- Fairly recent innovations allow *parallel processing*, doing more than one instruction at a time, e.g. the graphics processing units of Nvidia that significantly increase the computing speed for certain types of processing (in particular, neural networks).

- *Cloud computing services* such as Amazon Web Services, Microsoft Azure, and Google Cloud create huge computing centers that can take advantage of scale, allowing companies to use the most current technology over the Internet, charged on a usage basis without major capital expenditures. The generative AI services are also a form of cloud computing available to consumers and companies.
- Demands for computing power to develop ever-larger AI models have *increased the rate of investment* in computer center expansion. Mark Zuckerberg, Meta's boss, recently unveiled project Prometheus, a cluster of centers in Louisiana covering an area almost the size of Manhattan. According to *The Economist*, analysts at Morgan Stanley forecast $2.9 trillion will be spent on data centers and related infrastructure by the end of 2028; consultants at McKinsey put it at $6.7 trillion by 2030.
- *Quantum computers* are in an early stage but promise an eventual major jump in computing power for certain types of applications.

Digital systems have long driven changes in our lives. The Internet and World Wide Web are obvious examples, with related innovations including mobile phones, email, texting, and web search. Many of us spend a large fraction of the day working at our PCs using applications such as word processing and spreadsheets. But microprocessors also drive things we don't particularly think of as high technology, e.g. washing machines with their various choices of cycles. The declining cost of computer power will drive many innovations in all these applications.

Today's AI

The term 'artificial intelligence' is a bit misleading in that the technology driving what is called AI today is machine learning using deep neural networks ('deep learning'), a statistical technique that doesn't have as its core goal imitating human intelligence. It simply summarizes the implications of very large databases, although it uses a model that was indeed inspired by the way human neurons work. The technology can summarize databases too large for a human to even peruse, much less absorb, in a lifetime. I included in the chapter on AI examples of what the technology can do that goes well beyond human intelligence, e.g. a medical digital assistant that can tell which parts of cancer cells should be attacked to cure a cancer, a methodology leading to potential cures through genetic editing for *any* cancer. The core deep learning technology is very powerful.

Today, many consider the large language models of generative AI the next stage of AI and competitive with things such as writing articles associated with human intelligence. Generative AI still uses deep learning, but with very large neural nets. The assumption that a bigger neural net will be 'smarter' is challenged in this book and discussed briefly later in this chapter.

A promising step forward in neural nets is the mixture-of-experts model, where smaller specialized expert nets are trained and a supervisory neural net elicits their help. It has been shown to be less expensive and produce similar results to simply making a larger neural net.

Artificial General Intelligence

'Artificial general intelligence' (AGI) is supposedly the next stage of AI beyond what today's AI does. It is sometimes described as when AI equals human intelligence. The term 'superintelligence' is generally described as when AI *exceeds* human intelligence.

The goal of AI should be to be a useful tool to summarize the implications of large datasets, which it does with current technology. Since it does that by analyzing more data than a human could even examine in a lifetime, it has already exceeded human abilities. I've noted previously that, even at their earliest embodiment, computers exceeded human ability to do arithmetic quickly and accurately remember very large quantities of data.

Companies like OpenAI have stated their goal is to develop AGI. Karen Hao spent a lot of time interviewing executives at OpenAI for an article in *MIT Technology Review* in 2020. She described that experience and expanded her research in her 2025 book, *Empire of AI: Dreams and Nightmares in Sam Altman's OpenAI*. She tried to understand OpenAI's desire to develop AI and eventually AGI and reported general frustration in understanding what the company meant by AGI and how they interpreted that goal as consistent with the company's stated goal to make AI that benefited society. She summarized, "What's left unsaid is that in a vacuum of agreed-upon meaning, 'artificial intelligence' or 'artificial general intelligence' can be whatever OpenAI wants."

AGI is not a valid goal. It is certainly not a clear goal. To the degree it is a vague target to go beyond the current ability of deep learning, perhaps by improving on the mixture-of-experts approach, it may be a useful shorthand. But, as I have discussed in the subsection "The Data Problem" of the Artificial Intelligence chapter, just building larger neural nets may not produce the advance expected. I also noted that even the LLMs already developed are not being widely adopted by enterprises for operations because of the risks of unexpected and damaging results.

One could argue that the human brain is a large neural net, so a large simulation should in theory be able to compete with human intelligence. Experts have argued, however, that much of being 'human' depends on growing up in a human body; without a body, it is hard to fully understand the meaning of things like hunger or sex. I consider this a further argument that a goal of AGI should not depend on comparing it to humans.

Large Neural Networks

Will the very large neural nets being developed at great expense surprise us? Are they a threat to human existence through AGI or superintelligence?

I argued in the section on The Data Problem that the bigger-is-better theory runs into the problem of more parameters in a statistical model requiring exponentially more data to determine those parameters meaningfully. Using more parameters without more data can result in 'overfitting,' creating 'hallucinations' based on a few non-representative data points. Another issue is that just using all the data you can find forces the use of low-quality data.

Karen Hao reports in *Empire of AI* that OpenAI chose to expand the size of the neural net for GPT-3 despite a challenge finding enough data to drive the machine learning. For GPT-2, the company had been selective about what made it into the data. Hao said that sources told her for GPT-3 the company used a publicly available dataset known as Common Crawl, a database with millions of gigabytes of text scraped from all over the web—a source the company had previously avoided because it was such poor quality.

Growing even larger neural nets with GPT-4, the version available as this is written, further exacerbated the data problem. And the company is building an even bigger net with its Stargate project, as discussed. I've argued that the quality of good text data available today is overwhelmed by junk data full of misinformation and conspiracy theories, even for things like vaccines that shouldn't be political. Being forced to use even more data means the model is learning from mostly junk data. It will probably even learn to misspell words.

I also gave examples that suggested companies were having problems delivering the next version of their technology, even if it was not as ambitious as Stargate. There were announcements of upgrades that were repeatedly delayed. Further, companies trying to use even current versions were having practical problems due to the potential hallucinations the technology could generate. To illustrate possible problems, I suggested an LLM that gave medical advice could find itself facing a class action lawsuit by people taking harmful advice; even one such lawsuit with resulting publicity could discourage customer-facing use of the technology.

My prediction is that the bigger-is-better models will be a disaster. The possible exception was a switch from one big model to a collection of smaller models—the mixture-of-experts approach—possibly driven by independent 'clustering' of articles by their area of 'expertise.'

Accelerating growth in computer power

In the chapter on Computer Power, I provided a long list of ways that computer power was currently growing faster than Moore's Law predicted. Moore's Law described how the number of transistors you could fit on a chip would double every two years, but total computer power depended not only on the complexity of chips but also *how many chips* were manufactured and delivered. Further, chips that did computations in parallel, such as Nvidia's graphics processing units, did more computations at a time using those transistors. And cloud computing, huge computer centers such as Amazon Web Services, used economies of scale to deliver supercomputer power to any company on a usage basis.

There are more factors driving rapid growth in computer power, but the point is that we are in an era where computer power is expanding at an exponential rate, far beyond Moore's Law. The deep learning of today's AI is feasible, not because of some breakthrough in methodology, but because the cost of computing dropped to a point where the technology was economically feasible. With this example, it is fair to ask what that expanding resource will deliver next. It is fair to ask how those changes might affect society and the economy.

The economy

Paul Romer's classic paper, "Endogenous Technological Change," published in October 1990, distinguished "rival" goods (goods with a physical embodiment that limits sharing, e.g. oil) and "nonrival" goods (such as knowledge that can be stored in a form such as computer bits that allows, in principle, unlimited sharing). You can send a copy of a digital photo to a friend and still have it. Non-rival goods, such as a digital version of a hit song, aren't necessarily free, but the cost isn't a function of supply and demand. Endogenous entities such as digital assistants may impact economics significantly, for example, by changing the advertising market that has driven the success of companies like Google.

The cost of computer power is dropping at a fast rate. I've even argued that the failure of very big neural nets will result in computer centers intended to make such models becoming available for other uses. An oversupply of computer power will lead to a drop in its cost, allowing more applications that can use that power.

Historical innovations like electricity, automobiles, or the Internet have impacted us slowly by requiring a huge investment in infrastructure, giving the economy time to adjust. Computer power is expanding at a rate that makes slow adjustment difficult. It provides applications and services that cost almost nothing to deliver or update. As I discuss in The Next Big Thing subsection below, digital assistants may evolve such impressive abilities that they will change many aspects of technology and society, potentially making us smarter and less polarized.

Dangers

In the Artificial Intelligence chapter, I discussed agentic AI where AI, particularly generative AI, is given the power to take actions without supervision. In June 2025, *MIT Technology Review* quoted Yoshua Bengio, professor of computer science, University of Montreal, with a frightening warning on the subject: "If we continue on the current path...we are basically playing Russian roulette with humanity."

With appropriate instructions, I argued that a neural net could develop a sense of 'self' and 'self-preservation,' much as animals other than humans do. If the AI is 'agentic'—can take actions without human intervention—it is possible it might take actions that produce an unanticipated negative result to achieve its goals. To prevent being hacked, for example, it might launch a cyberattack on the hacker. In the extreme case, if the hacker is a country, the software might start a war. If this extreme sounds far-fetched to you, I must admit it does to me as well.

However, AI can cause a car to crash, and probably has. The Craft Law Firm analyzed data from the National Highway Traffic Safety Administration (NHTSA) database of vehicle incident reports. (Autonomous vehicle companies were required to report their crash data to the NHTSA as of June 2021.) The first reported incidents are from August 2019, and the data had been updated through June 17, 2024. There were a total of 3,979 incidents involving autonomous vehicles reported during this period. This data includes vehicles with automated driving systems and vehicles with advanced driver assistant systems. To put this number in perspective, in 2022, there were 42,514 motor vehicle crash *fatalities* in the United States; the large majority were of course the fault of humans. We try to reduce such terrible outcomes with technology such as seat belts. The point is to note that advances in technology have always come with potential dangers, not to minimize the impact on individuals and their families involved in an accident.

Another key point I have emphasized in this book is that deep learning is a *statistical* technique. It summarizes the data it is given to analyze; it can certainly give bad results if it learns from low-quality data or overfits the data it has because of using too many parameters—the 'bigger-is-better' fallacy. As with any statistical technique, it averages what it sees in the data to draw conclusions. Any average is not an exact prediction, but an attempt at a systematic prediction from history. The statistical nature of the technology is showing signs of limiting the practical application of the largest models, as discussed in the subsection on The Data Problem. In one example showing the limitations of a statistical model, in August 2025, a jury found Tesla partially liable for a fatal 2019 crash in Key Largo, Florida (the Tesla system supposedly required driver attention), and charged the company with $243 million in damages.

Of course, we must strive to find out if there are ways to minimize any danger from AI. But the reality is that such dangers have never stopped the growth of a technology that gave humans a useful new tool.

Talking to computers

Today, being able to use speech to interact with a digital system is widely available through digital assistants such as Apple's Siri, Google Assistant, and Amazon's Alexa. In an example of other speech recognition applications, you can dictate into Microsoft Word by simply clicking the 'dictate' icon. As someone who has worked on developing speech recognition technology, I am impressed with today's technology, and it seems users are beginning to take for granted this significant accomplishment. It is clearly something that is part of AI, since it mimics an ability once limited to humans.

Part of the breakthrough is the use of neural networks to identify the phonetic components of speech. The advance in capability of speech recognition, to repeat a theme, was made possible by the acceleration of computer power; you couldn't have a conversation with a computer if you had to wait several minutes for a response, for example.

A recent evolution is chatbots or large language models where the emphasis is typing a request for an explanation of a topic and being willing for the application to write a custom article which is in essence a summary of information in text sources. The functionality can be conceived of as a web search where the articles found are summarized into a relatively focused shorter article. This functionality is being characterized as an alternative to standard Web search producing a list of relevant websites, but doesn't make it easy to convey ads as does conventional web search. Even Google is adopting such a summary. Some of these chatbots and LLMs allow speech entry as well as text entry.

A recent evolution is chatbots or large language models where the emphasis is typing a request for an explanation of a topic and being willing for the application to write a custom article which is in essence a summary of information in text sources. The functionality can be conceived of as a web search where the articles found are summarized into a relatively focused shorter article. This functionality is being characterized as an alternative to standard Web search producing a list of relevant websites, but doesn't make it easy to convey ads as does conventional web search. Even Google is adopting such a summary. Some of these chatbots and LLMs allow speech entry as well as text entry.

The next big thing

I listed many predictions by experts over decades that our talking to digital assistants would be a major change in user interfaces to digital systems. Many even said it would make us smarter. But as this is written, digital assistants like Siri, Google Assistant, or Alexa are popular but have not become the dominant way of dealing with computers that was expected.

A key issue is the focus on speech as a sole interface. Privacy and politeness issues make it uncomfortable to talk to your digital device in public. A digital assistant you can't use in public is not one you want to depend on. The solution is currently available: almost all the major digital assistants now allow one to talk or type to them, and texting has made users comfortable in communicating by typing. The talk-or-text option eliminates the privacy and politeness issue. If there is a screen, it may also be useful when the response to a request requires a list or an image. In this case, the functionality might be called a 'talk, text, or touch' interface.

Large language models today are increasingly operating as a digital assistant, and the same consideration applies. Anthropic's Claude, for example, will accept both text and speech. The talk-or-text option hasn't been publicized by the big technology companies to date; I presume they consider it a beta test of the capability to simply introduce it without publicity.

Another issue with digital assistants has been responding to a request that maintains a dialog. Too often, the assistant simply says "Here's what I found" and defaults to a web search, dropping out of the conversation. Socrates criticized the written word because it was fixed; it couldn't maintain a dialog to explain itself. Today, LLMs are part of allowing a response—a discussion—a major improvement in digital assistants.

Another area is natural language processing to support speech or text requests. To be useful, a digital assistant has to *understand* a request to be able to generate a response or take an action. This aspect of language processing is also improving quickly, partly through neural nets.

Communicating with a digital system through human language is perhaps the most intuitive user interface. The talk-or-type option and mobile devices make the digital assistant always available. I've argued it almost becomes an extension of us, part of the evolution of what it means to be human. In any case, it makes us smarter, if only by giving us constant access to the intelligence stored on computers.

I'm not sure how fast digital assistants will become an indispensable tool. Their growth may be slowed if they become part of the polarization of US society, with people who don't trust the large technology companies accusing digital assistants of having a political agenda. If so, adoption may grow faster in other countries.

Further, a fully effective digital assistant requires work that is not yet evident in today's digital assistants. Switching between text and talk must be easy and obvious. The digital assistant needs to adapt seamlessly to the platform being used. The assistant must try to maintain a dialog and not automatically drop into a web search. Users must be aware of connections with companies through the assistant and how to use those connections. Security against hacking must be strong.

The large companies controlling the assistant will be motivated to avoid the assistants engaging in politically sensitive topics and motivated to protect the user from misinformation and objectional content. If the assistant becomes a major alternative news source and reduces interaction with the worse parts of social media, it might in the long term reduce polarization.

However, as I discussed, children from an early age could have an assistant that gives them a learning advantage and a 'friend.' If so, providing a child-oriented digital assistant to their progeny should eventually be a trend that is hard for parents to resist. I hope we don't have to wait a generation to see digital assistants reach their potential to improve human society.

15

References

Acemoglu, Daron, and James Robinson, *Why Nations Fail: The Origins of Power, Prosperity, and Poverty*, 2012.

Anthropic, "Agentic Misalignment: How LLMs could be insider threats", Jun 20, 2025, https://www.anthropic.com/research/agentic-misalignment.

Arthur, W. Brian, *The Nature of Technology: What It Is and How It Evolves*, 2009.

Atlantic Council, "What to know about foreign meddling in the US election," November 5, 2024.

Bellman, Richard, *Dynamic Programming*, 1957.

Baskaran, R., Abhishek Mahesh Appaji, and S. S. Iyengar, *Transforming Cybersecurity with AI A REVOLUTIONARY APPROACH*, IEEE, 2025.

Berlinski, David, *The Advent of the Algorithm: The 300-Year Journey from an Idea to the Computer*, 2001.

Bishop, Bill, *The Big Sort: Why the Clustering of Like-Minded America Is Tearing Us Apart*, 2008.

Blanchflower, David, , *Not Working: Where Have All the Good Jobs Gone?* 2019.

Breiman, Leo, Jerome Friedman, R.A. Olshen, Charles J. Stone, *Classification and Regression Trees,* 1984.

Bruni, Frank, *The Age of Grievance*, 2024.

Brynjolfsson, Erik and Andrew McFee, *The Second Machine Age*, 2016.

Calore, Raymond, *AI: The Great Disruption: How Artificial Intelligence Will Reshape Jobs, Society, and the Future of Humanity*, 2025.

Case, Anne and Deaton, Angus, *Deaths of Despair and the Future of Capitalism*, 2020.

Chen, Sophia, "IBM aims to build the world's first large-scale, error-corrected quantum computer by 2028," *MIT Technology Review*, June 10, 2025.

DeepSeek-AI, *DeepSeek-R1: Incentivizing Reasoning Capability in LLMs via Reinforcement Learning*, 2025.

Diamond, Jared, *The Third Chimpanzee: The Evolution and Future of the Human Animal*, 1992.

Eagleman, David, *The Brain: The Story of You*, 2015.

Emanuel, Rahm, "The Death of the American Dream" (interview), https://www.youtube.com/watch?v=hwymR8ZFdcs.

Ford, Martin, *Rule of the Robots: How Artificial Intelligence Will Transform Everything*, 2021.

Finn, Teaganne, and Downie, Amanda, "Agentic AI vs. generative AI," IBM, https://www.ibm.com/think/topics/agentic-ai-vs-generative-ai, 2025.

Gan, Guojun, Chaoqun Ma, Jianhong Wu, Data Clustering: Theory, Algorithms, and Applications, Second Edition, 2020.

Greene, Brian, *Until the End of Time: Mind, Matter, and Our Search for Meaning in an Evolving Universe*, 2020.

Grossberg, Stephen (editor), *Neural Networks and Natural Intelligence*, 1988.

Haass, Richard, *The Bill of Obligations*, 2023.

Hanania, Richard, *The Origins of Woke*, 2023.

Hao, Karen, *Empire of AI: Dreams and Nightmares in Sam Altman's OpenAI*, 2025.

Harman, Paul, and David King, *Expert Systems: Artificial Intelligence in Business*, 1985.

Hoffman, Reid, and Greg Beato, Superagency: What Could Possibly Go Right with Our AI Future, 2025.

IBM, "What is Unsupervised Learning?", https://www.ibm.com/think/topics/unsupervised-learning, 2021.

Isenberg, Nancy, *White Trash: The 400-Year Untold History of Class in America*, 2017.

Kahneman, Daniel, *Thinking Fast and Slow*, 2013.

Klein, Ezra, *Why We're Polarized*, 2020.

Korn Ferry Institute, The $8.5 Trillion Talent Shortage," https://www.kornferry.com/insights/this-week-in-leadership/talent-crunch-future-of-work.

Lee, Kai-Fu, *AI Superpowers: China, Silicon Valley, and the New World Order*, 2021.

Leonhardt, David, *Ours Was the Shining Future: The Story of the American Dream*, 2023.

Meisel, William, *Computer-Oriented Approaches to Pattern Recognition*, 1972.

Meisel, William, "The Personal Assistant Model: Unifying the Technology Experience," chapter in *Mobile Speech and Advanced Natural Language Solutions* (Amy Neustein and Dr. Judith Markowitz, Editors), Springer, 2013.

Meisel, William, *The Software Society: Cultural and Economic Impact*, 2013.

Meisel, William, *Evolution Continues: A Human-Computer Partnership*, 2022.

Meisel, William," The nature of time: Now is all that exists," https://medium.com/@billmeisel/the-nature-of-time-now-is-all-that-exists-78a4a73694c4 (Feb 2023).

Meisel, William, *The Lost History of "Talking to Computers": And What It Teaches Us About AI Exuberance*, 2025.

Miller, Chris, *Chip War: The Fight for the World's Most Critical Technology*, 2022.

Mohanty, Shayan, "The DeepSeek Series: A Technical Overview", https://martinfowler.com/articles/deepseek-papers.html, 2025.

Najjar, R., Redefining Radiology: A Review of Artificial Intelligence Integration in Medical Imaging, https://www.ncbi.nlm.nih.gov/pmc/articles/PMC10487271/, 2023.

O'Shea, Michael, *The Brain: A Very Short Introduction*, 2025.

OpenAI, *Announcing the Stargate Project*, https://openai.com/index/announcing-the-stargate-project/, January 21, 2025.

Peters U, Chin-Yee B., "Generalization bias in large language model summarization of scientific research," *R. Soc. Open Sci. 12: 241776*, 2025.

Rauch, Jonathan, *The Constitution of Knowledge*, 2021.

Romer, Paul, "Endogenous Technological Change," University of Chicago Press, October 1990.

Schmidt, Eric, interviewed by Bilawal Sidhu, May 15, 2025: *The AI Revolution Is Underhyped*, https://www.youtube.com/watch?v=id4YRO7G0wE.

Seth, Anil, *Being You*, 2021.

Schwartz, Eric Hal, "Dutch Study Examines the Benefits of Voice Assistants for Older People," February 26, 2020, https://voicebot.ai/2020/02/26/dutch-study-examines-the-benefits-of-voice-assistants-for-older-people/

Shojaee, Parshin, Iman Mirzadeh, Keivan Alizadeh, Maxwell Horton, Samy Bengio, Mehrdad Farajtabar, "The Illusion of Thinking: Understanding the Strengths and Limitations of Reasoning Models via the Lens of Problem Complexity", June 2025, https://machinelearning.apple.com/research/illusion-of-thinking.

Taylor, Jill Bolte, *My Stroke of Insight: A Brain Scientist's Personal Journey*, 2009.

The Economist, "Factory work is overrated. Here are the jobs of the future," June 10, 2025.

The Economist, "The world must escape the manufacturing delusion," June 13, 2025.

Topol, Eric, *Super Agers: An Evidence-Based Approach to Longevity*, 2025.

Tricks, Henry, "Welcome to the AI trough of disillusionment," *The Economist*, May 21, 2025.

Wolff, Michael, *Fire and Fury: Inside the Trump White House*, 2025.

Yang, Andrew, *The War on Normal People: The Truth About America's Disappearing Jobs and Why Universal Basic Income Is Our Future*, 2018.